1

Pre-Exam Fast Review

Disclaimer: All text, figures and ideas are my own. If there are any extreme similarities between this book and any publication written prior to the publication date of this book, I respectfully request the appropriate party to bring this to my attention and I will gladly remove the text or data if deemed appropriate. Best method of contact: email (ehabakkary@yahoo.com)

This book represents the ideas of the author and is not endorsed by the National Board of Medical Examiners (NBME). Please refer to www.usmle.org and the material provided to you by the NBME/USMLE® for any discrepancy, updates or changes.

Ehab Akkary MD FASMBS

ISBN 978-0-9839178-0-9

Published by Akkarypublishing LLC

Table of Contents

For my family, my beautiful daughter Talya and my mentor and professor, Dr. M. Gazayerli

I am indebted to all of you,
Ehab Akkary MD FASMBS

M. Gazayerli MD FRCSC
Bariatrics and Advanced Laparoscopy
Troy, MI

Why do you need this book?

- As a medical student seeking residency, you need to demonstrate certain skills that are assessed by the USMLE® step 2 CS exam.

- My aim, through this book, is to help you approach the cases encountered during this exam in a simple systematic way, obtain the necessary clinical information, perform a relevant physical exam and conclude with the appropriate differential diagnosis and workup.

- The step 2 CS exam is a great test to help you think in a well-organized fashion that you can apply thereafter during your practice as a resident or in your career as a physician.

- When I was a Medical Student, I was told by one of my professors "If you follow a systematic approach, you will never miss a diagnosis".

- Remember, you always want to be the doctor who orders the echocardiogram after auscultating the heart not the one who's auscultating the heart because there is a valve lesion on the echo!!

- It is true that recognition memory is significantly stronger than recall memory. Based on this fact, this book is designed to make it extremely easy and simple for you to remember all the important points needed to be covered during the exam using diagrams, mnemonics and a standard approach that you can implement in all cases.

- When I took the Step 2 CS exam (formerly CSA), I thought it was the easiest test in the entire USMLE® series.

- However, what I heard from other students before the exam was worrisome; "It took me 3 months to study for this test", "Time is never enough", "I didn't know what to ask the patient", "Some cases are very difficult and I couldn't reach a differential diagnosis".

- Now, let me share with you what I thought about this exam; "It took me 1 week to study for it", "Time is almost always enough but might be tight", "I never had a problem reaching the differential diagnosis and workup", "All cases are straightforward".

- Before I took the exam, I searched the web looking for feedback from students who failed the test and their comments were very similar; "not enough time", "don't know what to ask"…etc.

- To avoid this, I am introducing in this book the concept of the "**Keyword**".

- You will find the "**Keyword**" in the Chief Complaint. Based on the "**Keyword**" you will decide on what questions to ask and which physical exam to perform. This book should help you develop a systematic plan based on the "**Keyword**" in order to approach the cases in a methodical, organized and timely fashion.

The Exam

- The CS exam intends to evaluate your communication as well as clinical skills.

- The patients you will encounter are not real patients but in fact your examiners!! However, don't think of them as "standardized patients or examiners", think of them as patients and of yourself as the physician, this mindset and their excellence in "acting" will help you believe they are real patients rather than examiners.

- Recognizing the "**Keyword**" is the most important initial step. The first question in your mind once you see the patient is *"What system am I examining?"*

- <u>Very important</u>: The "**Keyword**" originates from the Chief Complaint.

- The patient will state a complaint → specify which system is being examined → Perform the history and physical exam (H&P) of this system.

- Example: Patient with right upper quadrant pain would have an H&P similar to a patient with left lower quadrant pain (both are "*gastrointestinal*" cases). Focus on examining <u>SYSTEMS</u> rather than isolated complaints.

- The complaint represents the "**Keyword**" that leads you to the <u>SYSTEM</u>.

- Once you specify the system, then ask all the questions related to this system without being distracted by the complaint; e.g. a case of hand numbness and a case of stroke represent 2 "**keywords**" that lead you to the "*Neuro*" system so your H&P will be that of the Neuro system virtually regardless of the complaint.

- Some students think of the cases as individual complaints then they try to memorize H&Ps based on specific complaints, for example, they consider a case of vomiting and another of diarrhea as separate cases and study each individually instead of thinking of both as **keywords** to the same "*gastrointestinal*" system.

- Don't ask in medical terms e.g. ask about "shortness of breath" not "dyspnea", ask about "heart racing or awareness of heartbeats" not "palpitations or arrhythmia".

· <u>Follow this general algorithm for all cases:</u>

A] <u>Outside the room:</u>
1) At the door → read the chief complaint → specify your "**Keyword**" and therefore the <u>SYSTEM</u> you are targeting (e.g. GI, chest, cardiac, neuro, etc…)

2) Write the vital signs on your scratch paper.

3) Knock on the door.

B] <u>Inside the room (you are allowed 15 minutes):</u>
1) Introduce yourself "e.g. good morning Mr. (last name), my name is Dr. (last name)". Give a confident handshake and proper eye contact.

2) Wash your hands before the physical exam. Don't examine through the patient's gown. Don't repeat maneuvers unnecessarily (especially if painful).

3) Always tell the patient what you are going to do e.g. "now Mr./Ms ….. I need to examine you, would you mind if I untie your gown?", "now I am going to listen to your heart", "I am going to push on your belly".

4) After the "**Keyword**" H&P → explain to the patient your differential diagnosis and diagnostic workup in a simple way (e.g. don't say "CBC with differential", say "blood work") and address the patient's concerns (e.g. ask if the patient needs a pain medication or if he/she has any questions). Feel free to hand a crying patient a tissue from over the sink, this demonstrates your compassion.

C] <u>Outside the room again (10 minutes):</u>
1) Write or type your "**Keyword**" H&P.

2) Give 5 differential diagnoses.

3) Give 5 diagnostic workups.

When you type or write your note, you are allowed to use common abbreviations. This might save you plenty of time.

1) Year-old: YO
2) Male: M
3) Female: F
4) African American: AA
6) Caucasian: C
7) Left: L
8) Right: R
9) History of: h/o
10) Complaining of: c/o
11) History of present illness: HPI
12) Past medical history: PMH
13) Past surgical history: PSH
14) Medications: Meds
15) No known drug allergy: NKDA
16) Social history: SH
17) Family history: FH
18) Obstetrics and gynecology: OB/GYN
19) Last menstrual period: LMP
20) Cigarettes: cig
21) Alcohol: ETOH
22) Intravenous drug abuse: IVDA
23) Congestive heart failure: CHF
24) Chronic obstructive pulmonary disease: COPD
25) Hypertension: HTN
26) Coronary artery disease: CAD
27) Myocardial infarction: MI
28) Coronary artery bypass grafting: CABG
29) Jugular venous distention: JVD
30) Normal sinus rhythm: NSR
31) Diabetes Mellitus: DM
32) Transient ischemic attack: TIA
33) Cerebrovascular accident: CVA
34) Upper respiratory tract infection: URI
35) Within normal limits: WNL
36) No abnormality detected: NAD
37) Without or no: Ø
38) Positive: +ve

39) Negative: -ve
40) Head, eyes, ears, nose, and throat: HEENT
41) Ears, nose, and throat: ENT
42) Pupils are round, regular and reactive: Pupils are RRR
43) Abdomen: Abd
44) Gastrointestinal: GI
45) Genitourinary: GU
46) Extremities: Ext
47) Deep tendon reflexes: DTR
48) Blood urea nitrogen: BUN
49) Creatinine: Cr
50) Complete blood count: CBC
51) Red blood cells: RBC
52) White blood cells: WBC
53) Hemoglobin: Hgb
54) Hematocrit: Hct
55) Platelets: Plt
56) Prothrombin time: PT
57) Partial thromboplastin time: PTT
58) Urinalysis: U/A
59) Chest X-ray: CXR
60) Computed tomography: CT scan
61) Magnetic resonance imaging: MRI
62) Electrocardiogram: EKG
63) Nothing by mouth: NPO
64) Intramuscular: IM
65) Intravenous: IV
66) Lumbar puncture: LP
67) Right upper quadrant: RUQ
68) Right lower quadrant: RLQ
69) Left upper quadrant: LUQ
70) Left lower quadrant: LLQ

· Remember, This exam is nothing more than what Medical Students do everyday all around the world.

· Please note that the **history** components are the same for all patients *except* history of present illness (HPI).

- Remember, the first question in your mind once you obtain the Chief Complaint (which represents the "**Keyword**") is:_"What SYSTEM am I examining?"_

- Then ask all the questions listed under that system and perform the relevant physical exam.

- In your H&P, list the points covered under that system whether positive or negative.

- This approach might make you ask a few more questions than what you need but will save you a lot of time in the thought process and will help you stay focused.

HISTORY

1) Chief Complaint (CC)

(In patient's own words) → "What troubles you?" "What brings you here today?" + duration e.g. "shortness of breath for 2 days". (THE COMPLAINT IS YOUR **KEYWORD** TO THE SYSTEM YOU ARE EXAMINING)

2) History of Present Illness (HPI): = analysis of complaint + specific "Keyword".

Analysis of complaint:
[A] Onset (sudden or gradual?)
[B] Course (Is it getting better, worse or same?)
[C] Duration (for how long?).

Specific "Keyword": will be discussed in detail according to individual systems.

3) Past Medical History (PMH):

"Do you have any medical problems?" You can give examples to the patient like diabetes, high blood pressure, asthma.

4) Past Surgical History (PSH):

"Did you have any surgeries before?" You can give examples like appendix, gall bladder.

5) Past OB/GYN History:

(a) If the complain doesn't belong to the "female genital system", ask: "Any problem with the cycles and when was the LMP?"

(b) If the complain belongs to the "female genital system", ask the questions listed under that system.

6) Medications:

"Are you taking any medications?"

7) Allergy:

"Do you have any allergies?"

8) Social History:

Ask about history of smoking, alcohol, IVDA

9) Social History:

Ask about any medical problems in the family.

Important:

- Don't forget to ask about the "**General**" symptoms with any **System**

- GENERAL symptoms are "constitutional": (1) Fever/chills
  (2) Weight $\downarrow$ or $\uparrow$
  (3) Appetite change

- Pain is a very common complaint, analyzing the pain properly might lead you to the diagnosis.

- **ANY PAIN** = 4 questions(1) Onset
  (2) Location & radiation
  (3) Quality & severity
  (4) What $\downarrow$ or $\uparrow$

- In this book, the systems are listed as mnemonics and simple sketch diagrams to facilitate your recall.

- Analyze the chief complaint **"Keyword"** in terms of onset, course, duration then the rest of your HPI should include the symptoms listed under every system.

- For example; 47 yo male w diarrhea → the **Keyword** is diarrhea then the SYSTEM is "gastrointestinal" → remember the HPI is the only variable among the different cases → so you need to ask about onset/course/duration of diarrhea → then you ask all the other symptoms listed under that system e.g. difficult or painful swallowing, nausea, vomiting, peptic ulcer, abdominal pain, pancreatitis, hemorrhoids, bleeding per rectum, etc...... If the patient mentioned abdominal pain → this should trigger 4 questions immediately: *(1) onset (2) location & radiation (3) quality & severity (4) what $\downarrow$ or $\uparrow$.* Don't forget the "**General**" symptoms with any system you are examining (see above).

- Always remember that female, with abdominal pain, in childbearing age maybe **PREGNANT.** A pregnancy test should always be included in your workup.

PHYSICAL EXAM

- The physical exam commonly includes inspection, palpation, percussion, auscultation.

- Please remember that we added "<u>General</u>" to all systems during history taking (General was the constitutional symptoms).

- We will follow the same pattern in the physical exam and add "General Exam" to all cases, this will be: <u>Auscultate heart & chest</u>.

- This does not have to be a detailed heart and chest examination when the focus of the case is on another system.

- So you are quickly screening the heart and chest for an abnormality while you are examining another system e.g. GI in detail.

- For example, after you auscultate the heart and chest, if you do not find an abnormality, you can say:

  Heart is RRR (i.e. regular rate and rhythm),

  Chest is CTA (B) (i.e. clear to auscultation bilaterally)

- When you write (or type) your physical exam, you can follow this general format:

  - Patient is alert, oriented (unless noted otherwise e.g. delirium case)
  - Vitals: T ___, P ___, BP ___, R ___ (you will be provided with the vitals before you go into the room)
  - Heart: RRR
  - Chest: CTA (B)
  - The specific system findings

Diagrams and Mnemonics

o In this section of the book, I use simple sketch diagrams and mnemonics to facilitate your recall.

o The aim is to eliminate the time you might lose trying to think of what questions to ask or what signs to look for in the physical exam.

o Diagrams and mnemonics will be listed for the history and physical exam of each system.

o Use the following chart to follow a systematic approach.

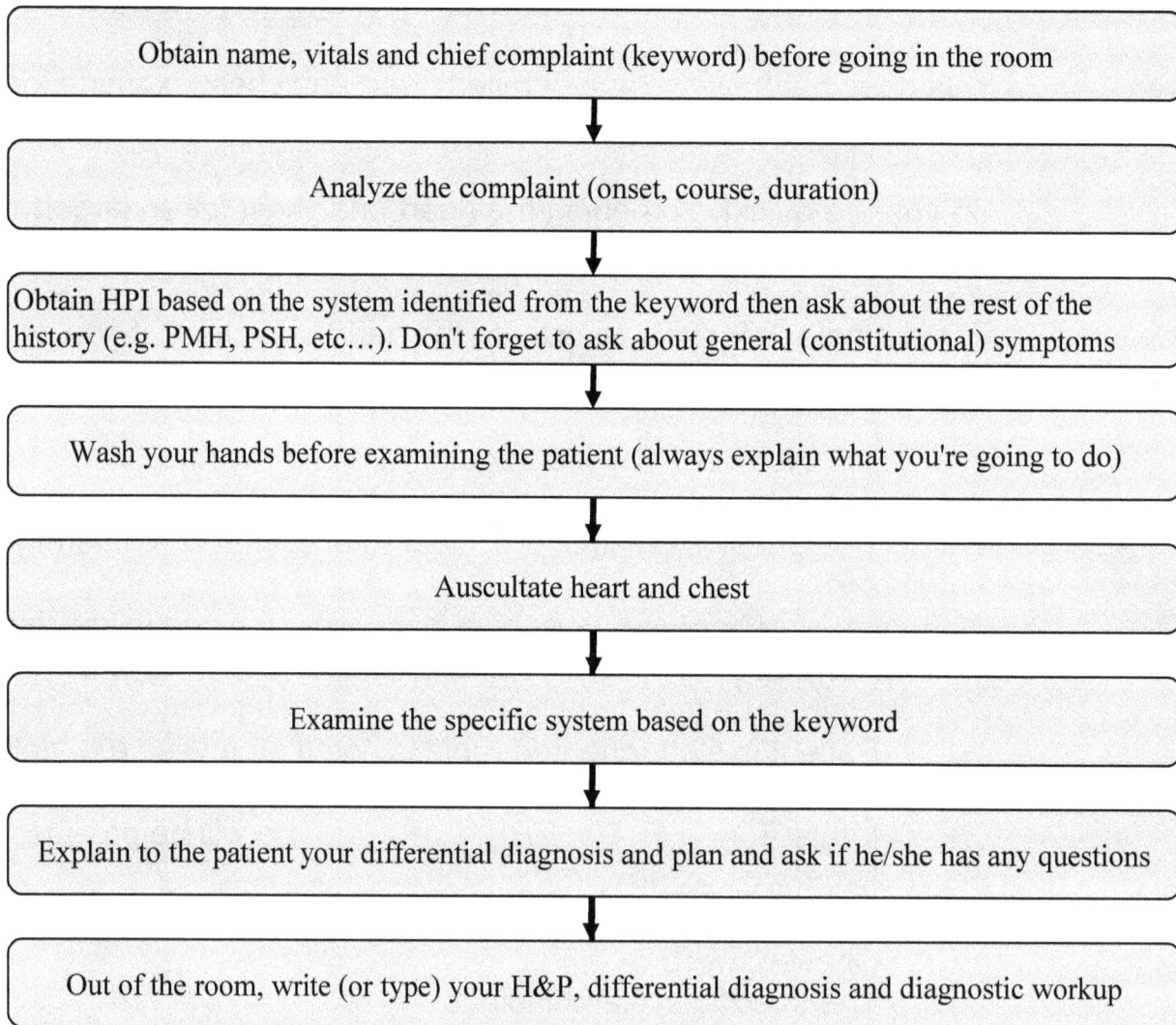

o

Obtain name, vitals and chief complaint (keyword) before going in the room

↓

Analyze the complaint (onset, course, duration)

↓

Obtain HPI based on the system identified from the keyword then ask about the rest of the history (e.g. PMH, PSH, etc…). Don't forget to ask about general (constitutional) symptoms

↓

Wash your hands before examining the patient (always explain what you're going to do)

↓

Auscultate heart and chest

↓

Examine the specific system based on the keyword

↓

Explain to the patient your differential diagnosis and plan and ask if he/she has any questions

↓

Out of the room, write (or type) your H&P, differential diagnosis and diagnostic workup

GASTROINTESTINAL (GI) HISTORY

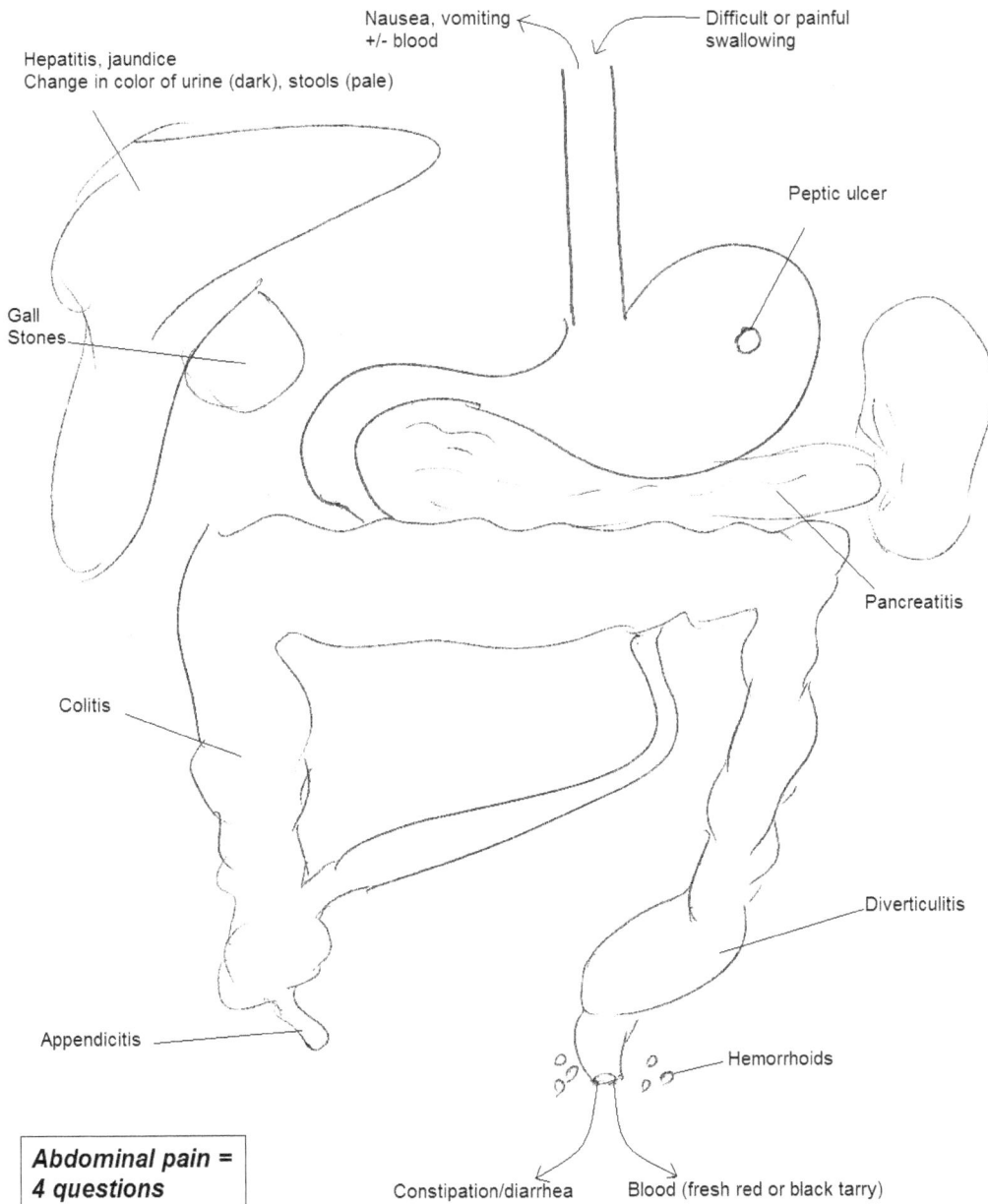

Hepatitis, jaundice
Change in color of urine (dark), stools (pale)

Nausea, vomiting
+/- blood

Difficult or painful
swallowing

Peptic ulcer

Gall
Stones

Pancreatitis

Colitis

Diverticulitis

Appendicitis

Hemorrhoids

**Abdominal pain =
4 questions**

Constipation/diarrhea

Blood (fresh red or black tarry)

** VERY IMPORTANT: Any female in childbearing age → Ask about the possibility of PREGNANCY & sexually transmitted disease.

GASTROINTESTINAL (GI) EXAM

(a) Inspection: scars, swelling, distension, dilated veins.
(b) Auscultation: bowel sounds (normal, hypo or hyperactive).
(c) Percussion: in all 4 quadrants (look for dullness e.g. hepatomegaly).
(d) Palpation: superficial then deep while looking at patient's face for tenderness. Tap the flank for CVA tenderness

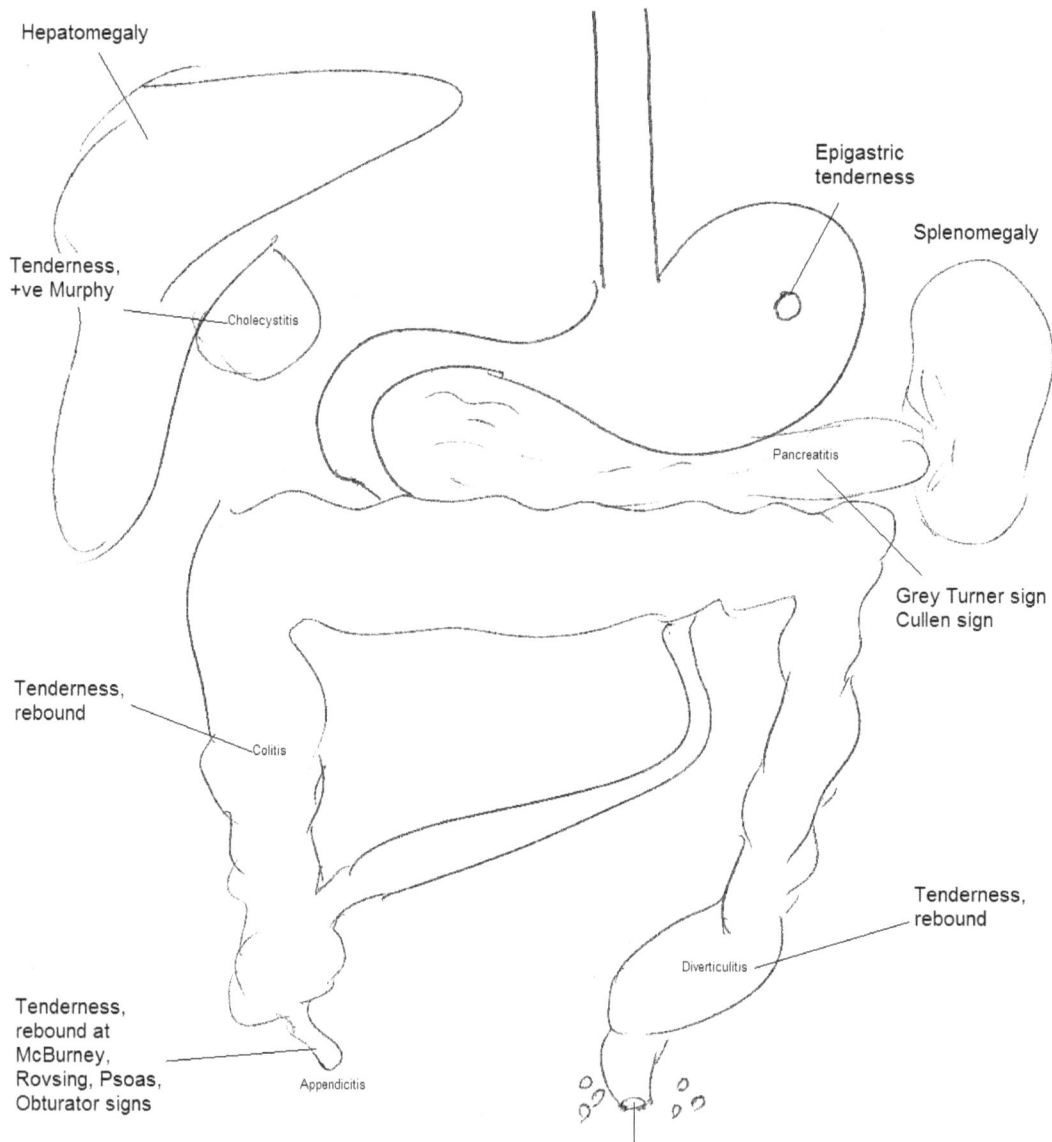

Hepatomegaly

Epigastric tenderness

Splenomegaly

Tenderness, +ve Murphy

Cholecystitis

Pancreatitis

Grey Turner sign
Cullen sign

Tenderness, rebound

Colitis

Tenderness, rebound

Diverticulitis

Tenderness, rebound at McBurney, Rovsing, Psoas, Obturator signs

Appendicitis

Remember you are not allowed to perform rectal exam but you can include it in your diagnostic workup

CHEST HISTORY

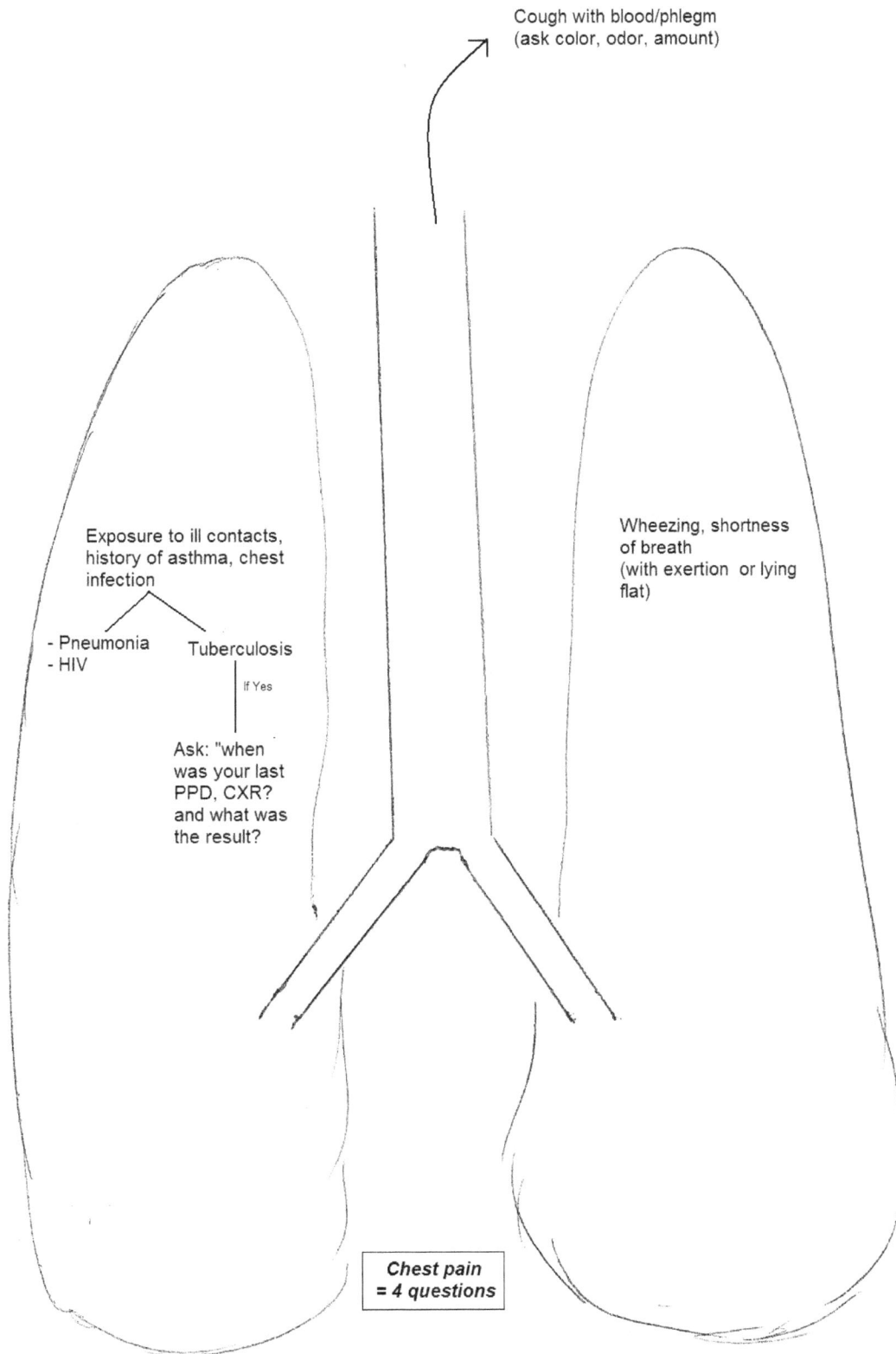

Cough with blood/phlegm
(ask color, odor, amount)

Exposure to ill contacts,
history of asthma, chest
infection

- Pneumonia
- HIV

Tuberculosis

If Yes

Ask: "when
was your last
PPD, CXR?
and what was
the result?

Wheezing, shortness
of breath
(with exertion or lying
flat)

*Chest pain
= 4 questions*

CHEST EXAM

Have the patient in sitting position

(a) *Inspection*: breathing pattern, asymmetry, using accessory muscles

(b) *Palpation*: **3T** (**T**racheal deviation, **T**enderness, **T**actile vocal fremitus).

(c) *Percussion*: from lung apex to base, front & back.

(d) *Auscultation*: remember **PECT**oralis

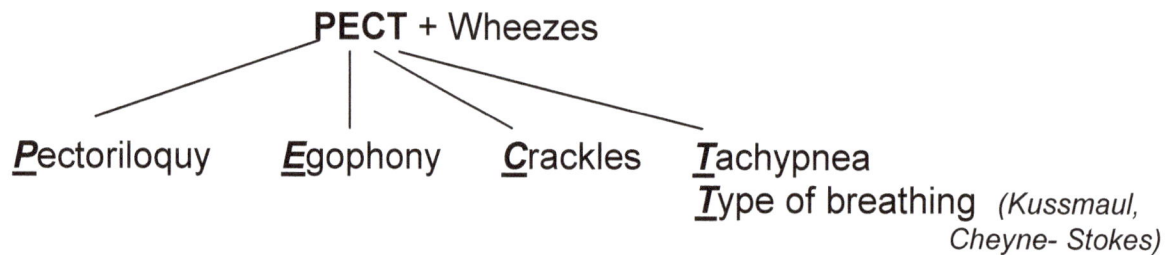

PECT + Wheezes

Pectoriloquy   **E**gophony   **C**rackles   **T**achypnea
                                                        Type of breathing *(Kussmaul, Cheyne- Stokes)*

CARDIAC HISTORY
(A P C D + chest pain)

A → Angina or infarction.
P → Palpitations, Pressure (retrosternal pressure – high blood pressure)
C → CHF, Cholesterol
D → Dizziness, Dyspnea ...ask about shortness of breath +/- (1) exertion
(2) at night
(3) laying flat

+ Chest pain (4 questions)

CARDIAC EXAM

Remember the cardiac exam includes neck, liver and legs to look for congestion signs indicating heart failure.

Have the patient lying down with head elevated at 30°:
(a) *Inspection*: JVD (jugular venous distension), PMI (point of maximal impulse)
(b) *Palpation*: Thrills over precordium, lower extremity edema
(c) *Percussion*: of the RUQ for hepatomegaly.
(c) *Auscultation*: 4 auscultatory areas looking for murmurs, rubs, gallops

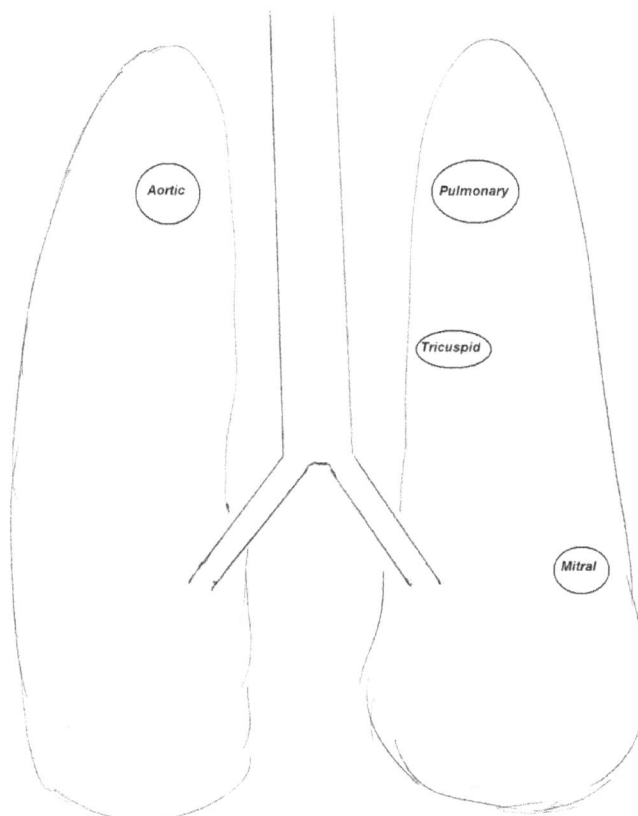

NEUROLOGICAL HISTORY

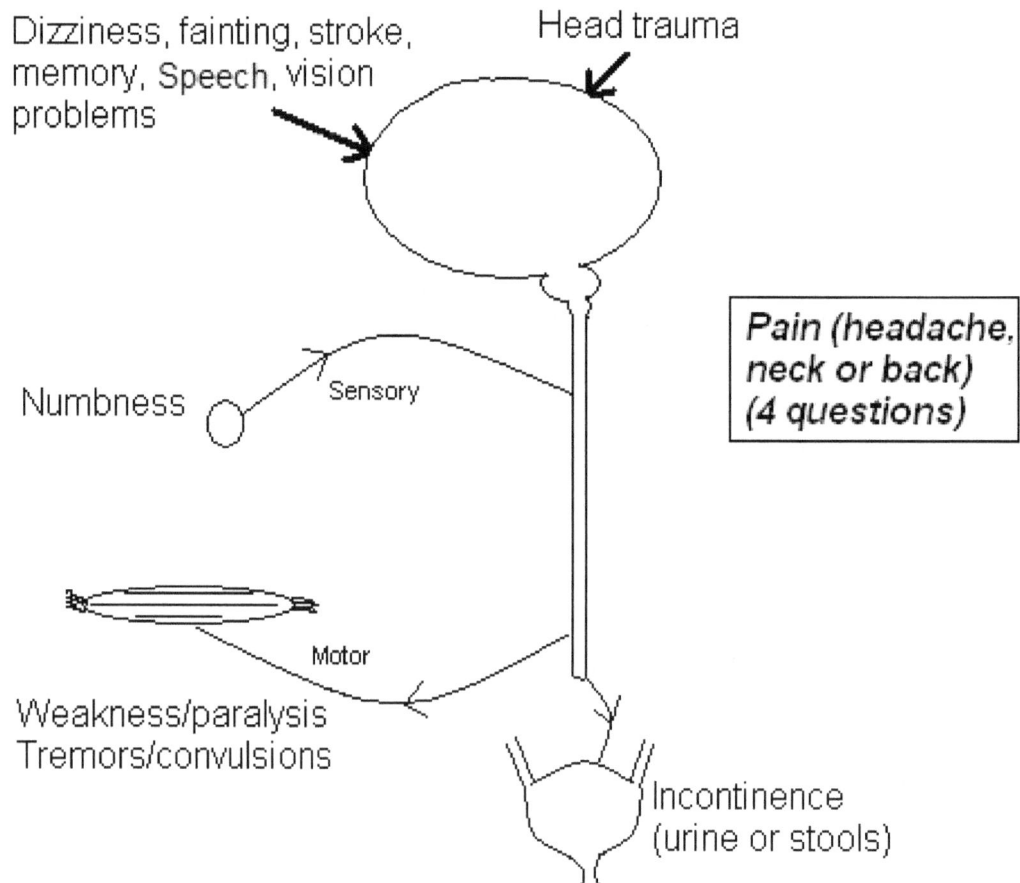

Dizziness, fainting, stroke, memory, Speech, vision problems

Head trauma

Numbness

Sensory

Pain (headache, neck or back) (4 questions)

Motor

Weakness/paralysis
Tremors/convulsions

Incontinence (urine or stools)

NEUROLOGICAL EXAM (compare the 2 sides)

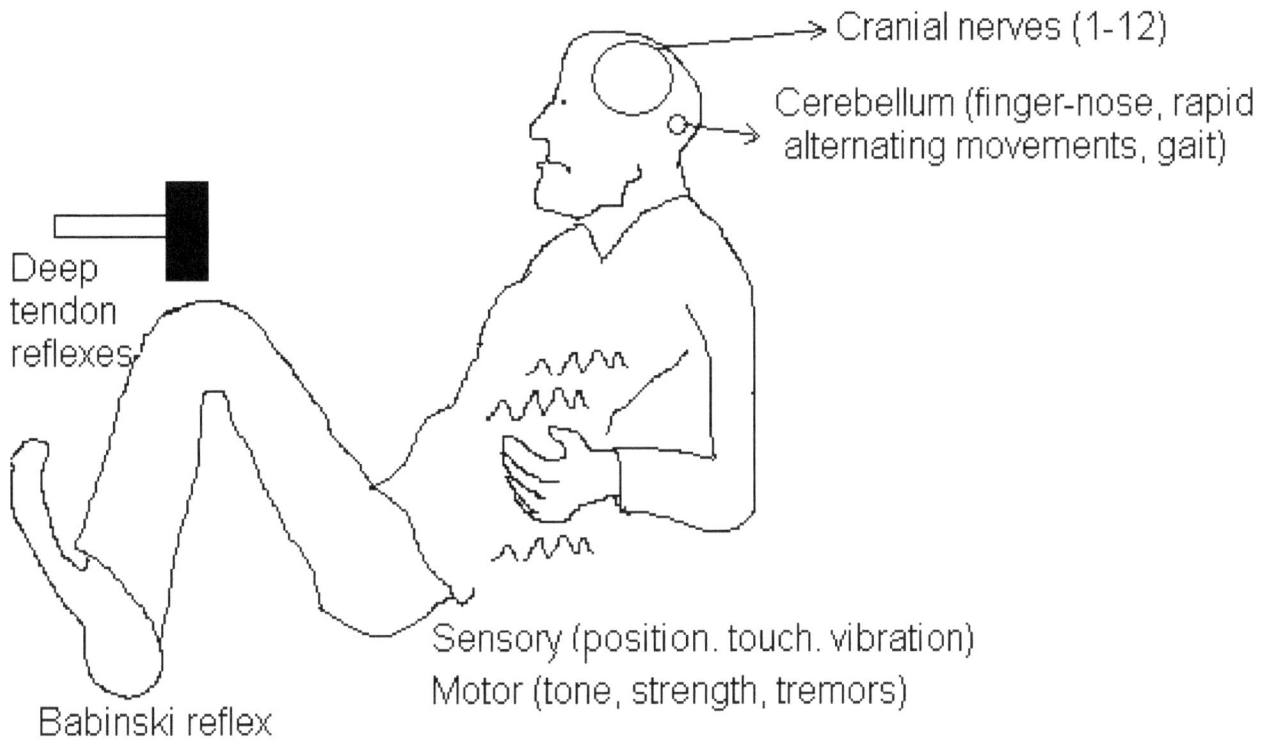

Cranial nerves (1-12)

Cerebellum (finger-nose, rapid alternating movements, gait)

Deep tendon reflexes

Babinski reflex

Sensory (position. touch. vibration)
Motor (tone, strength, tremors)

BREAST HISTORY

Lumps & family
history of breast
cancer

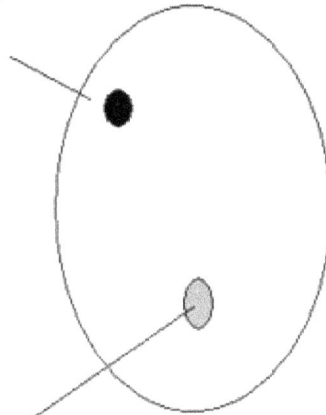

Nipple retraction,
discharge, bleeding

**Breast pain =
4 questions**

+ MAM

1) mammogram
2) assymetry
3) monthly self exam

BREAST EXAM

not done but should be included in the diagnostic workup.

EYES HISTORY

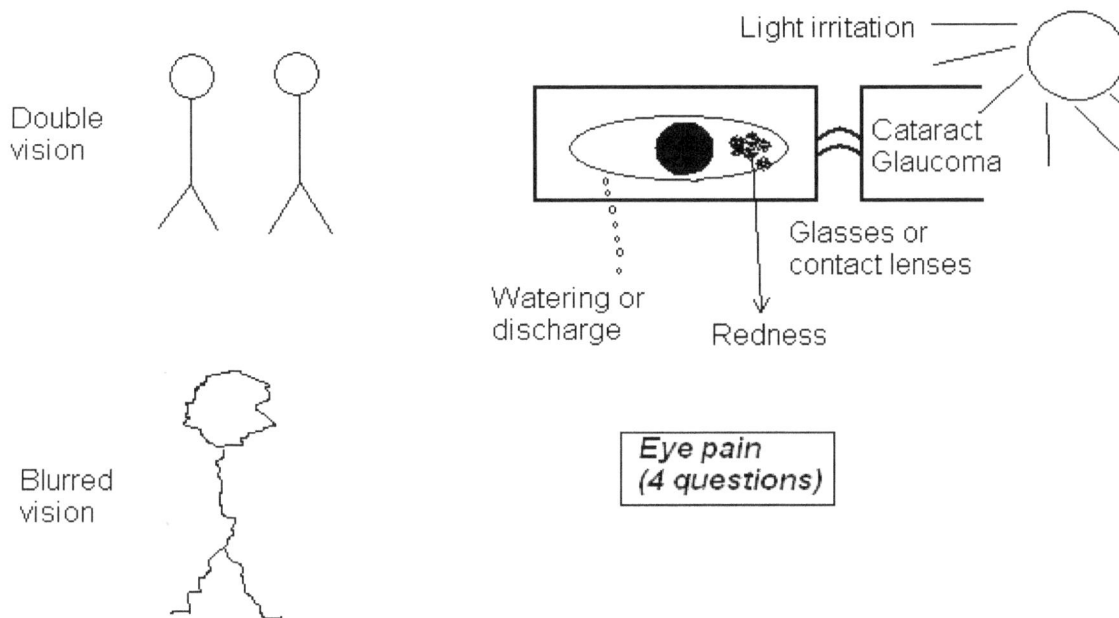

Double vision

Blurred vision

Light irritation

Cataract
Glaucoma

Watering or discharge

Glasses or contact lenses

Redness

Eye pain
(4 questions)

EYES EXAM (compare the 2 sides)

Use the ophthalmoscope and comment on the following:

1- Eye lids: stye, chalazion, ptosis, ectropion
2- Conjunctiva: discharge, erythema
3- Cornea: abrasion (usually Fluorescein test is needed which you will not perform in the exam but you can include it in your diagnostic workup)
4- Pupils: RRR (round, regular and reactive)

Beyond that you need different equipment that you can include in your diagnostic workup. For example, slit lamp to examine the lens, fundoscope to examine the fundus.

EARS HISTORY

Decreased hearing
(uni/bilateral)
(partial/complete)

Dizziness &
vertigo

Trauma
(noise or
physical)

Ringing

Discharge &
infection

Ear pain
(4 questions)

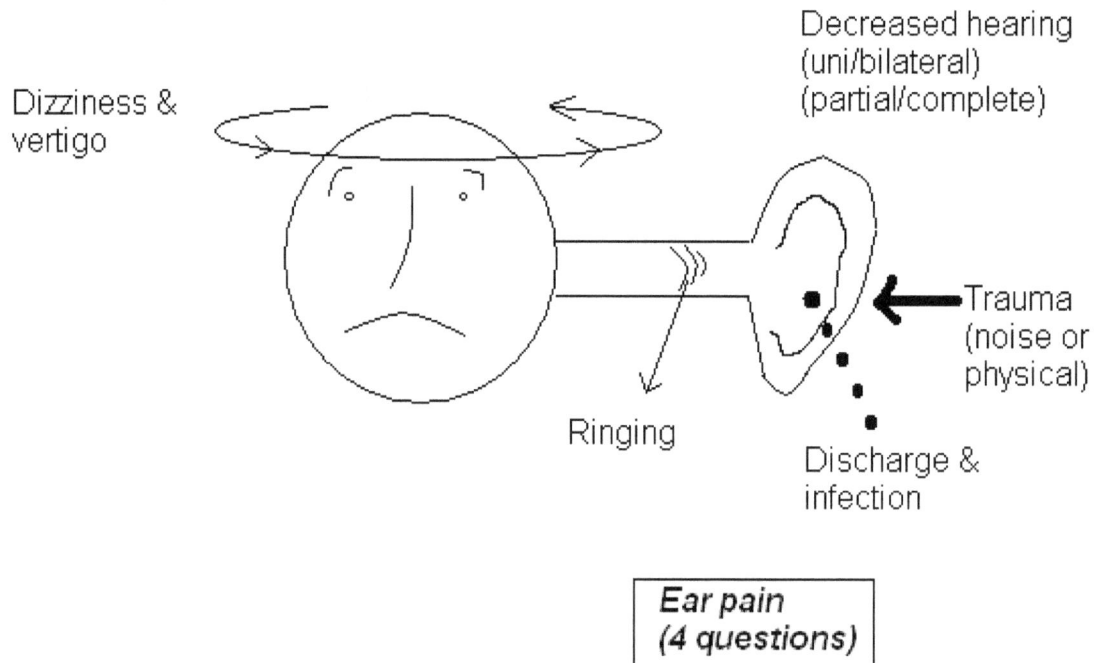

EARS EXAM (compare the 2 sides)

Otoscope + 2 tests (Weber & Rinne).

THROAT HISTORY

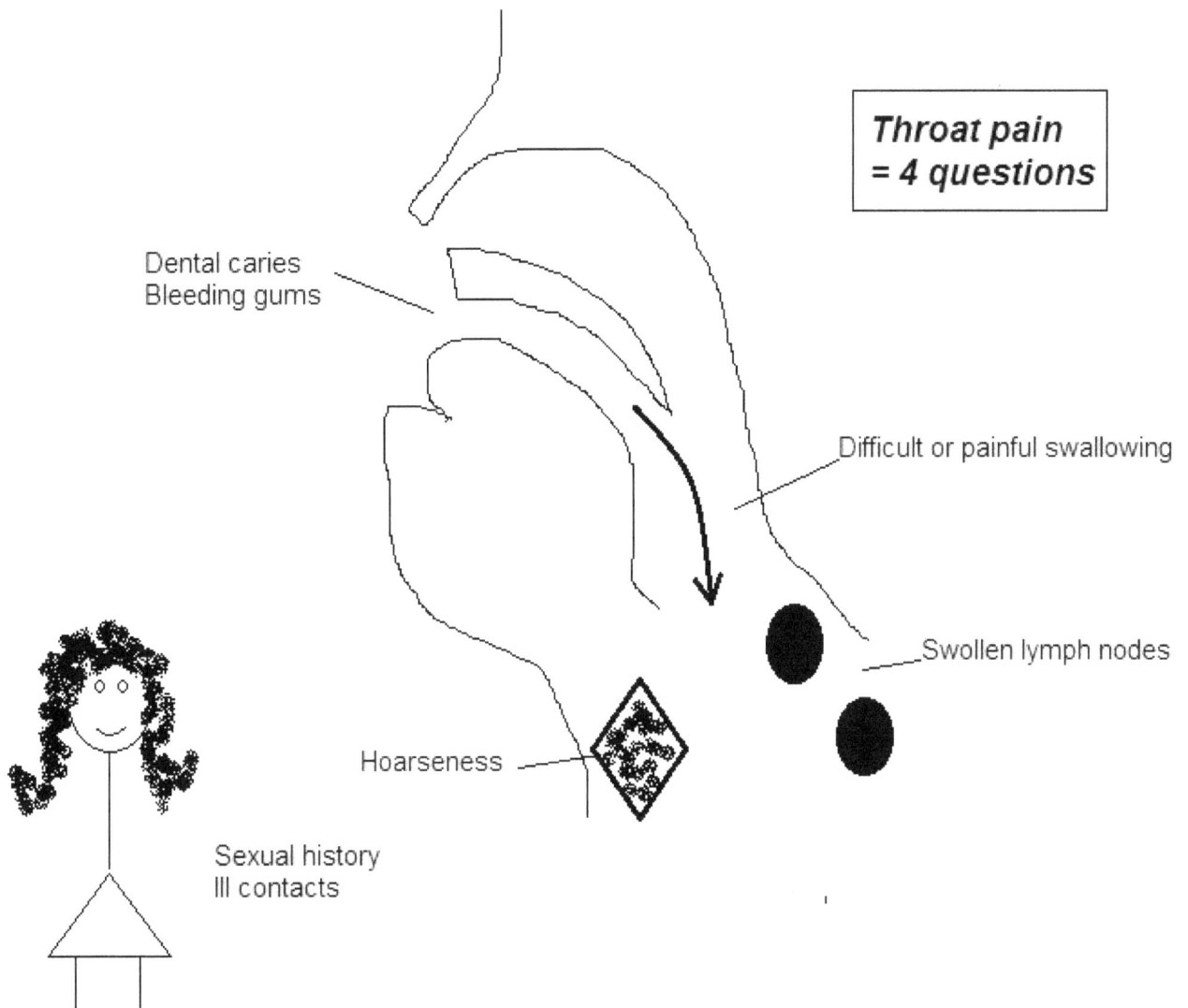

Throat pain
= 4 questions

Dental caries
Bleeding gums

Difficult or painful swallowing

Swollen lymph nodes

Hoarseness

Sexual history
Ill contacts

THROAT EXAM

(a) *Inspection*: ask the patient to open mouth to examine the pharynx with light source, ask the patient to swallow to look for goiter moving with deglutition.
(b) *Palpation*: thyroid, cervical lymph nodes

ENDOCRINE HISTORY

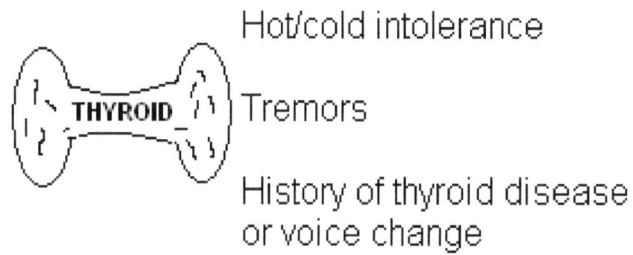

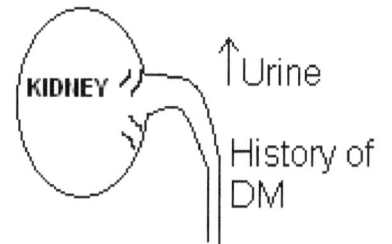

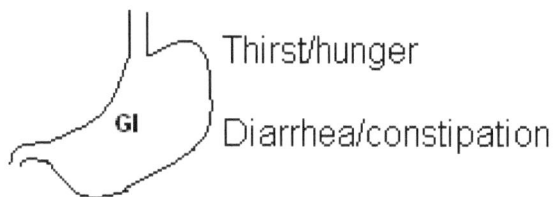

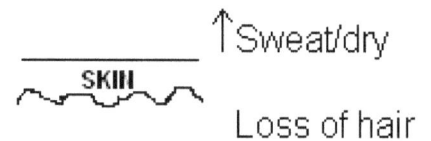

Hot/cold intolerance

THYROID

Tremors

History of thyroid disease or voice change

KIDNEY

↑Urine

History of DM

GI

Thirst/hunger

Diarrhea/constipation

SKIN

↑Sweat/dry

Loss of hair

ENDOCRINE EXAM

- *Palpate* thyroid for goiter, nodules.
- *Palpate* sternal notch for tracheal deviation.
- *Palpate* lymph nodes (submental, submandibular, cervical).

PSYCHIATRY HISTORY

(= *SADNESS*)

S → Sad

A → Anxiety

D → Depressed, hopeless

N → Nervous, irritable

E → Ear (hearing voices)

S → Sexual, Sleep problems

S → Suicidal ideation, Support system

PSYCHIATRY EXAM

MAD x 3 = (3M + 3A + 3D)

3M → **M**ania – **M**ania (hypo) – **M**ental status change (minimental exam).

3A → **A**nxiety – **A**ttack (panic) – **A**nger.

3D → **D**ementia – **D**epression – **D**elirium.

MUSCULOSKELETAL HISTORY
(The system to run FAST !!)

F → Fractures (ask about mechanism, abuse).

A → Arthritis (osteo, rheumatoid, gout).

S → Swelling or deformity of joint.

T → Trauma

Joint or back pain = 4 questions

MUSCULOSKELETAL EXAM (compare the 2 sides)

a) *Inspection*: swelling, erythema, contusion
b) *Palpation*: tenderness, ↓ range of motion, joint deformity
c) **Adson test** (for thoracic outlet syndrome)
d) **Phalen & Tinel tests** (for carpal tunnel syndrome)

VASCULAR HISTORY

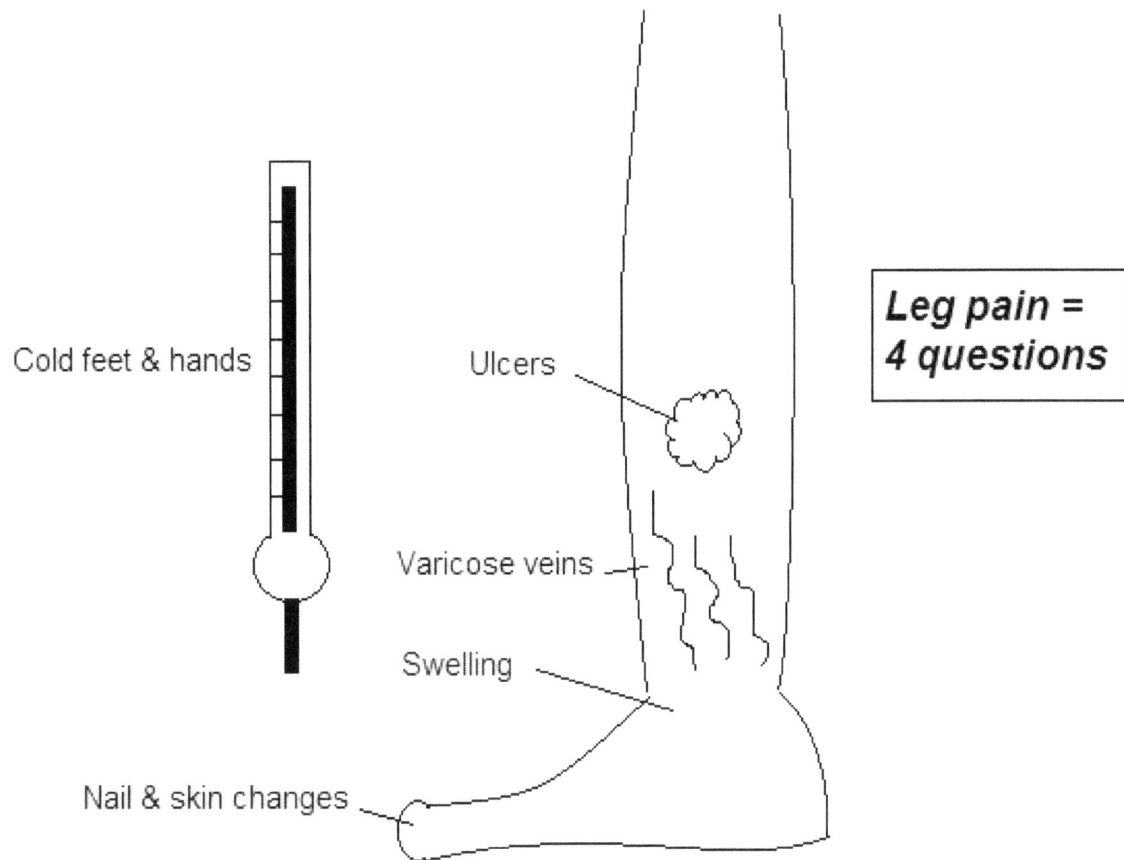

Cold feet & hands

Ulcers

Varicose veins

Swelling

Nail & skin changes

Leg pain = 4 questions

Don't forget to ask about shortness of breath & chest pain for pulmonary embolism

VASCULAR EXAM (compare the 2 sides & assess arterial/venous)

[A] Arterial signs:

1) Peripheral pulses (radial, femoral, popliteal, posterior tibial, dorsalis pedis).

2) Loss of hair, clubbing, cyanosis

[B] Venous signs:

1) Edema, varicose veins.

2) Calf tenderness, Homan's sign.

UROLOGY HISTORY

Use the mnemonic: **"NO STONE IN BLADDER"**

NO → **NO**cturia

STone → **ST**ones, **ST**D, **ST**ream (weak stream, difficult urination)

IN → **IN**continence, **IN**complete emptying, **IN**fection

Bladder → **B**urning with urination, **B**loody or cloudy urine

Flank pain = 4 questions

UROLOGY EXAM

- Same as GI exam (don't forget to tap the flank for CVA tenderness).

- Mention to the patient that part of the diagnostic workup would be genital examination (remember you are not allowed to perform that but you should include it in the diagnostic workup).

- Older male: also mention rectal exam as part of your diagnostic workup to check the prostate

MALE GENITAL SYSTEM HISTORY

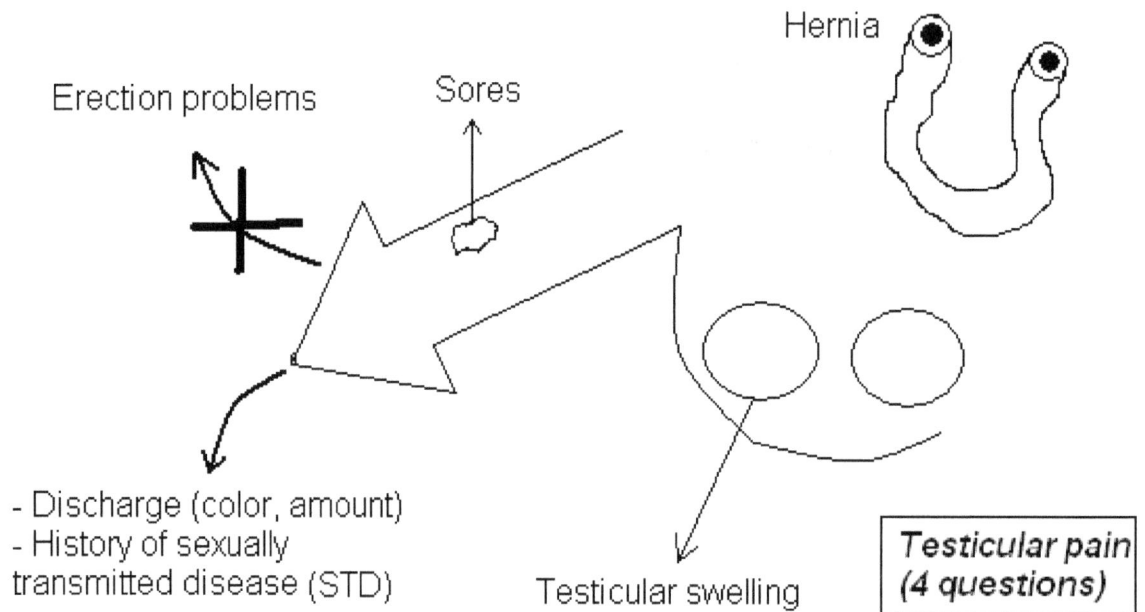

Erection problems

Sores

Hernia

- Discharge (color, amount)
- History of sexually transmitted disease (STD)

Testicular swelling

Testicular pain (4 questions)

MALE GENITAL SYSTEM EXAM

Not done but should be included in the diagnostic workup.

FEMALE GENITAL SYSTEM HISTORY

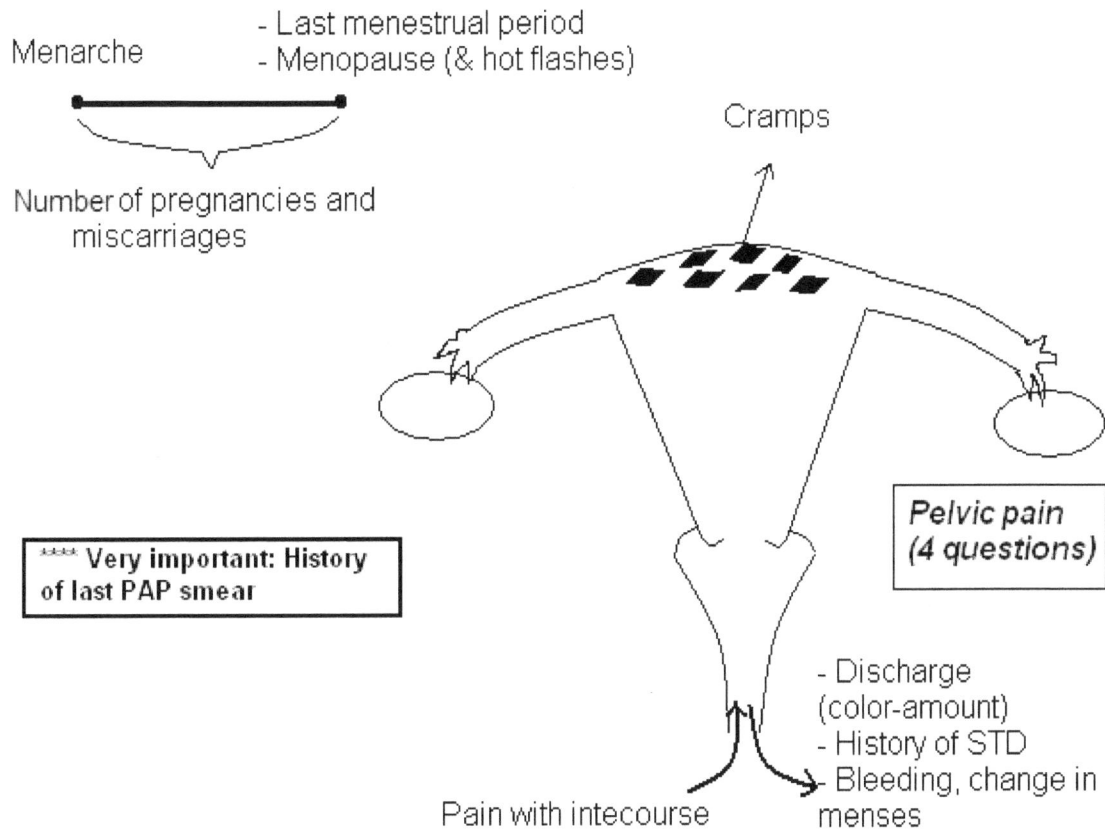

Menarche

- Last menestrual period
- Menopause (& hot flashes)

Number of pregnancies and miscarriages

Cramps

**** Very important: History of last PAP smear

Pelvic pain (4 questions)

- Discharge (color-amount)
- History of STD
- Bleeding, change in menses

Pain with intecourse

FEMALE GENITAL SYSTEM EXAM

Not done but should be included in the diagnostic workup.

Clinical Cases

- o The case starts with a "complaint" which represents your keyword → Identify the system based on the keyword.

- o Don't forget to include "General" in your history which is the constitutional symptoms.

- o Don't forget to include "General" in your physical exam which is auscultating heart and chest.

- o Below you will find multiple chief complaints (keywords), the keyword = the system you are being tested on + "general".

- o You need to include 5 differential diagnoses (DD) and 5 diagnostic workups (WU).

- o Remember that breast, pelvic, genital and rectal exams are not done but are included in the diagnostic workup.

- o The listed DD and WU do not have to exactly correspond e.g. (#1 in WU does not necessarily mean that it represents a test for #1 in the DD).

- o Small sized diagrams are inserted for your recall, please refer to the full size ones under the appropriate sections.

[1] Right lower abdominal pain in male = GI + general

DD	WU
1- Acute appendicitis	1- CBC with differential
2- Ureteral colic	2- Urine analysis
3- Inflammatory bowel disease (Ulcerative colitis, Crohn's)	3- CT scan of the abdomen & pelvis
4- Gastroenteritis	4- Abdominal X-ray (supine – erect)
5- Inguinal hernia	5- Colonoscopy, rectal exam & hemoccult

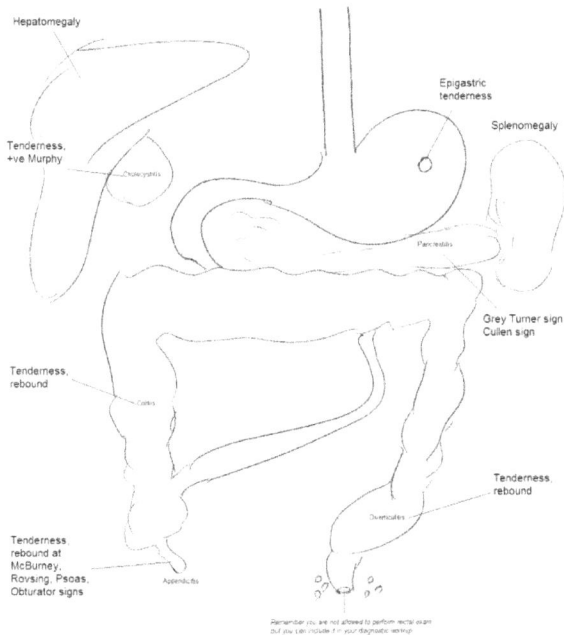

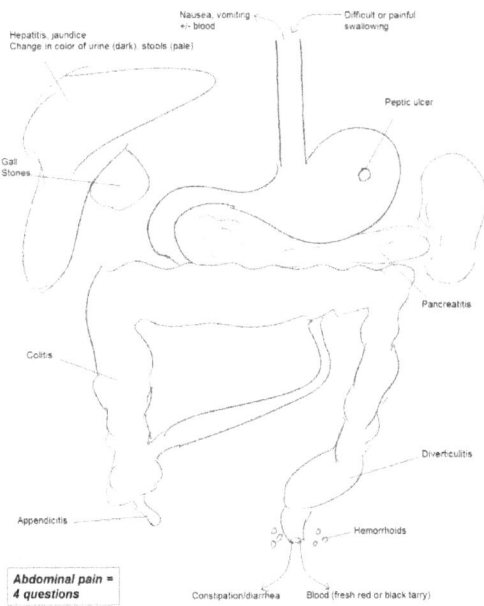

[2] Left lower abdominal pain in male = GI + general

DD	WU
1- Acute diverticulitis	1- CBC with differential
2- Ureteral colic	2- Urine analysis
3- Inflammatory bowel disease (Ulcerative colitis, Crohn's)	3- CT scan of the abdomen & pelvis
4- Gastroenteritis	4- Abdominal X-ray (supine – erect)
5- Inguinal hernia	5- Colonoscopy, rectal exam & hemoccult

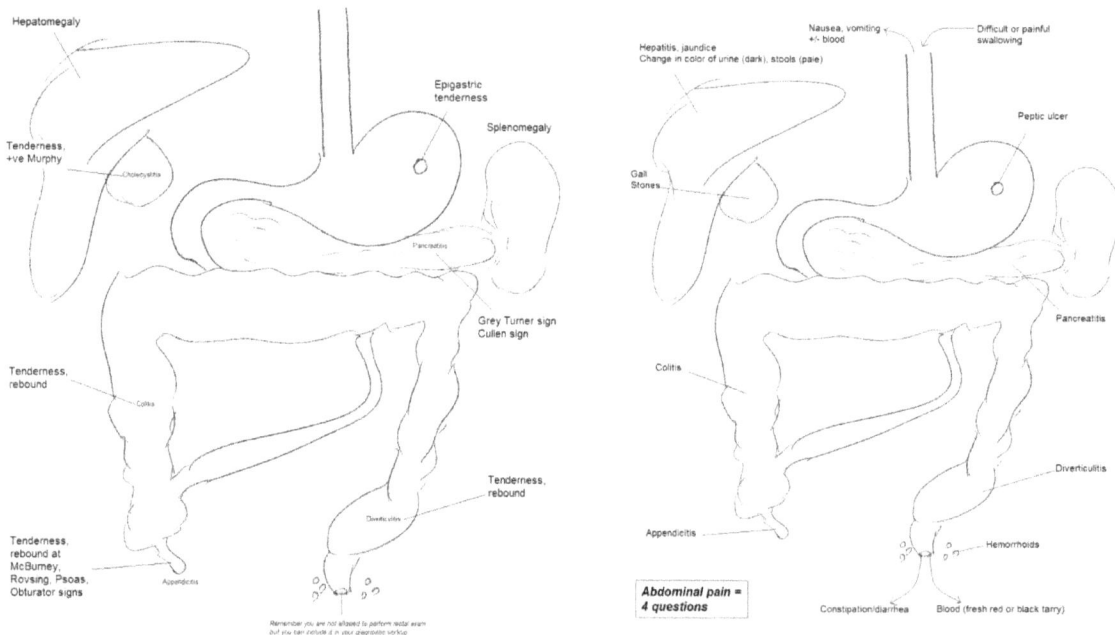

[3] Right lower abdominal pain in female = GI + female genital system + general

DD	WU
1- Acute appendicitis	1- CBC with differential
2- Ureteral colic	2- Urine analysis
3- Inflammatory bowel disease (Ulcerative colitis, Crohn's)	3- **Serum & urine β-HCG level**
4- Ovarian torsion	4- Abdominal & pelvic ultrasound
5- Ectopic pregnancy	5- CT scan of the abdomen & pelvis
6- Pelvic inflammatory disease	6- Abdominal X-ray (supine – erect)
7- Endometriosis	7- Vaginal culture eg Chlamydia, gonorrhea
8- Gastroenteritis	8- Colonoscopy
9- Inguinal hernia	9- Rectal exam & hemoccult

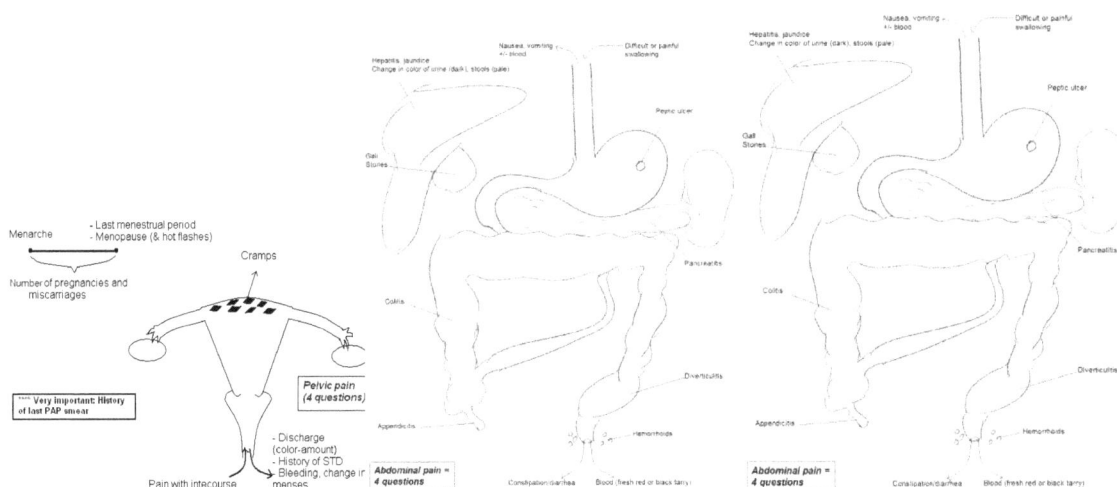

[4] Left lower abdominal pain in female = GI + female genital system + general

DD	WU
1- Acute diverticulitis	1- CBC with differential
2- Ureteral colic	2- Urine analysis
3- Inflammatory bowel disease (Ulcerative colitis, Crohn's)	3- **Serum & urine β-HCG level**
4- Ovarian torsion	4- Abdominal & pelvic ultrasound
5- Ectopic pregnancy	5- CT scan of the abdomen & pelvis
6- Pelvic inflammatory disease	6- Abdominal X-ray (supine – erect)
7- Endometriosis	7- Vaginal culture eg Chlamydia, gonorrhea
8- Gastroenteritis	8- Colonoscopy
9- Inguinal hernia	9- Rectal exam & hemoccult

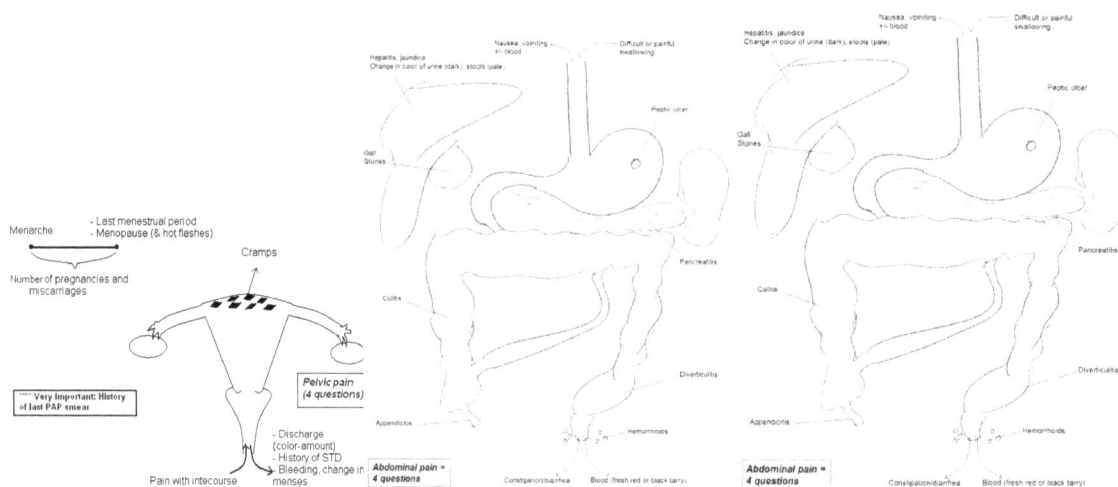

[5] Right upper abdominal pain (male or female) = GI + general

DD	WU
1- Cholecystitis	1- CBC with differential
2- Hepatitis	2- Liver function tests
3- Right lower lobe pneumonia	3- **Serum & urine β-HCG level** in female
4- Colitis	4- Abdominal ultrasound
5- Cardiac (angina – infarct)	5- 12 lead EKG, cardiac enzymes
6- Pyelonephritis	6- CXR (PA- Lat)
7- Peptic ulcer	7- Urine analysis
8- Musculoskeletal	8- CT scan of the abdomen
	9- Upper GI endoscopy

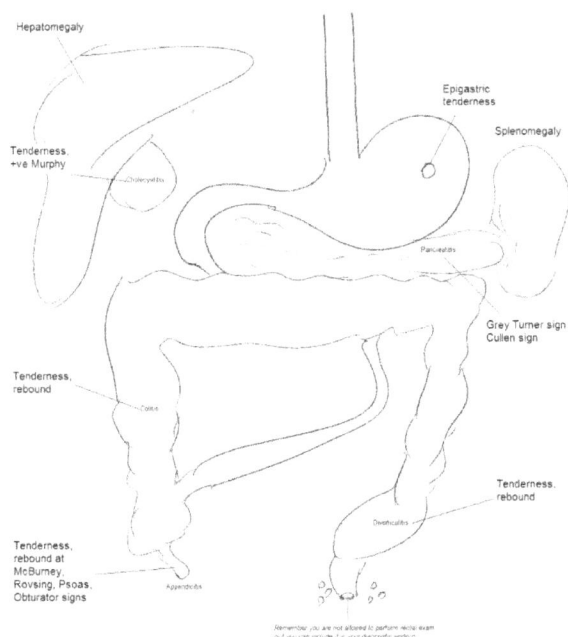

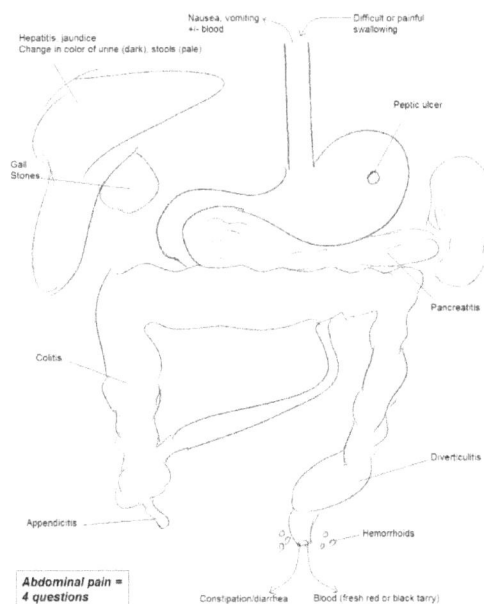

[6] Left upper abdominal pain (male or female) = GI + general

DD	WU
1- Splenic infarct, splenomegaly	1- CBC with differential
2- Peptic ulcer	2- 12 lead EKG, cardiac enzymes
3- Left lower lobe pneumonia	3- **Serum & urine β-HCG level** in female
4- Colitis	4- Abdominal ultrasound
5- Cardiac (angina – infarct)	5- CXR (PA- Lat)
6- Pyelonephritis	6- Upper GI endoscopy
7- Musculoskeletal	7- Urine analysis
	8- CT scan of the abdomen

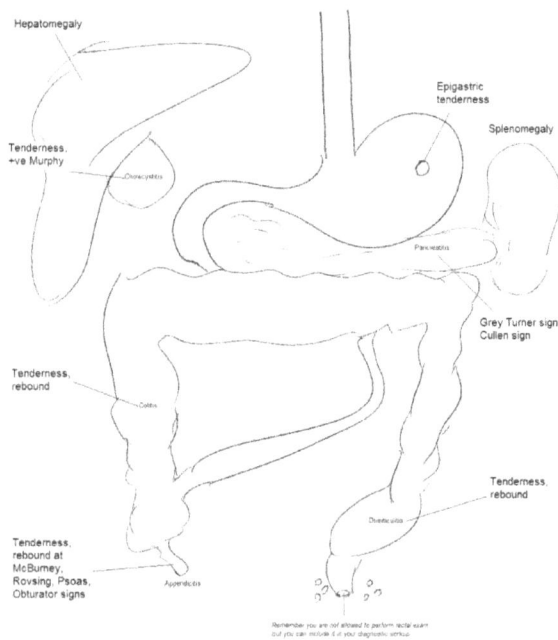

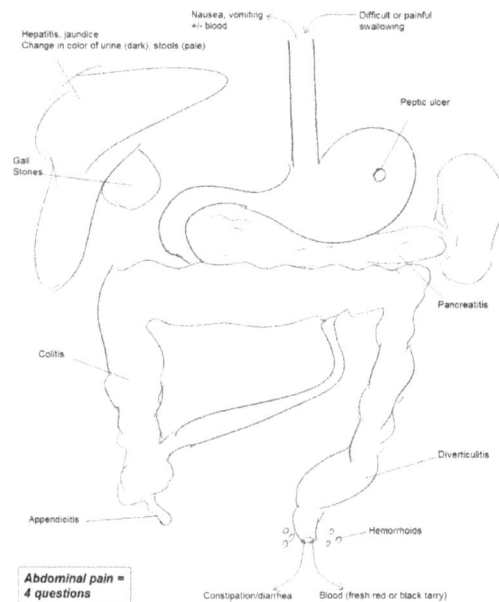

[7] Epigastric pain (male or female) = GI + general

DD	WU
1- GERD	1- CBC with differential
2- Peptic ulcer disease	2- Upper GI endoscopy
3- Pancreatitis	3- **Serum & urine β-HCG level** in female
4- Cardiac (angina – infarct)	4- CT scan of the abdomen
5- Gastritis	5- 12 lead EKG, cardiac enzymes
6- Pancreatic cancer	6- Serum amylase, lipase
7- Crohn's disease	7- CA 19-9 tumor marker

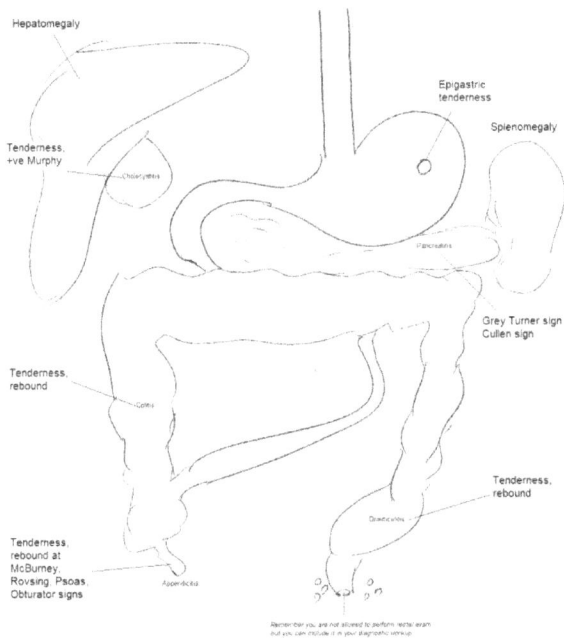

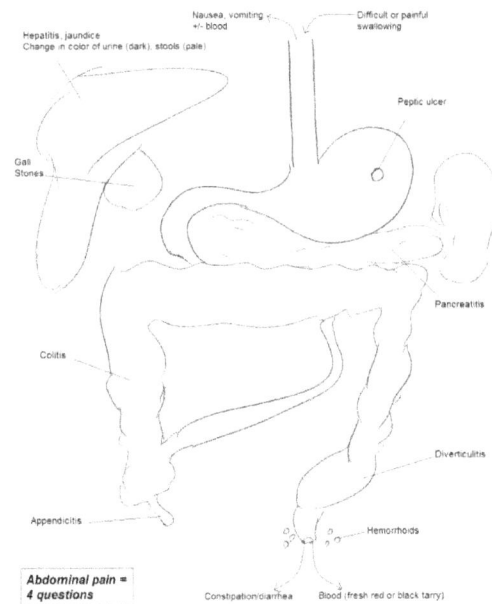

[8] Vomiting blood = GI + general

DD	WU
1- Esophageal varices	1- CBC with differential
2- Peptic ulcer disease	2- Serum electrolytes
3- Mallory-Weiss tear	3- Upper GI endoscopy
4- Gastric cancer	4- Serum PT, PTT, INR
5- Dieulafoy's lesion	5- Angiography
6- Trauma, retching	6- Urea breath test
6- Aortoenteric fistula	7- Tagged RBC scan

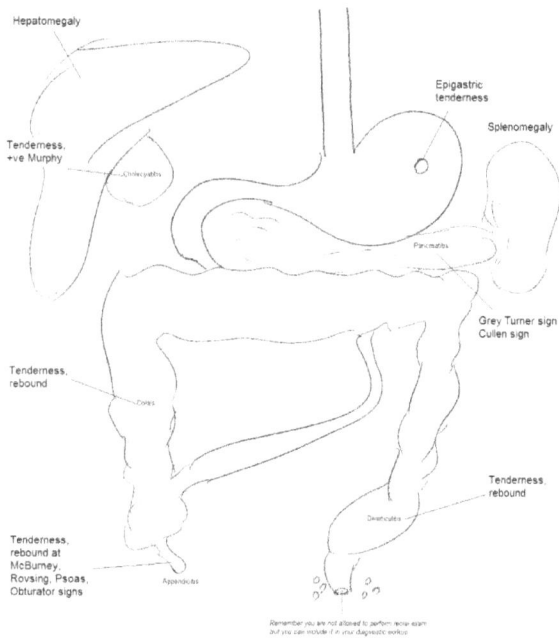

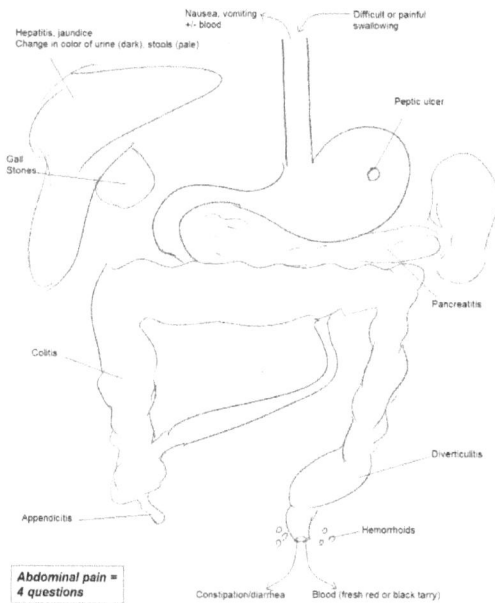

[9] Bleeding per rectum = GI + general

DD	WU
1- Diverticulosis	1- CBC with differential
2- Angiodysplasia	2- serum electrolytes
3- Inflammatory bowel disease (Ulcerative colitis, Crohn's)	3- Rectal exam
4- Infectious colitis	4- Colonoscopy
5- Pseudomembranous colitis	5- Stools for C-Diff toxin, ova, parasites, G stain, C&S
6- Ischemic colitis	6- Tagged RBC scan
7- Hemorrhoids	7- Mesenteric angiography
9- Colorectal polyps/cancer	8- Serum CEA level

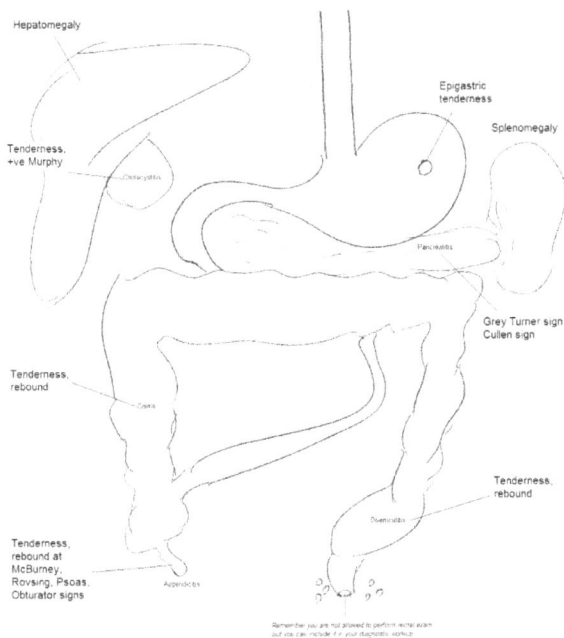

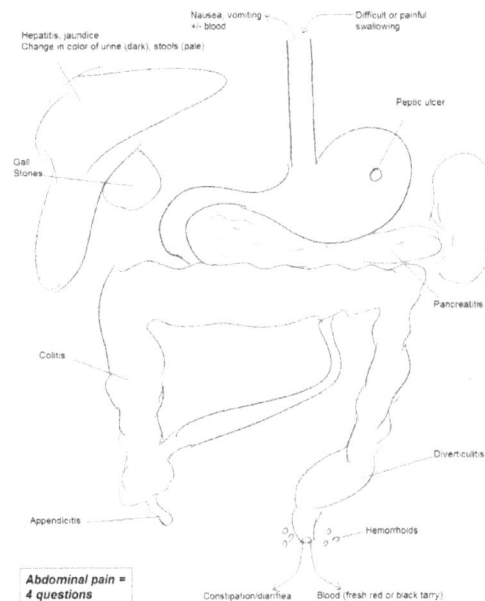

[10] Yellowish eye or skin discoloration = GI + general

DD	WU
1- CBD stones	1- CBC with differential
2- Biliary stricture	2- serum electrolytes
3- Tumor in the head of the pancreas	3- Abdominal Ultrasound
4- Hepatitis	4- CT scan of the abdomen
5- Hemolytic anemia	5- Liver function tests
	6- ERCP

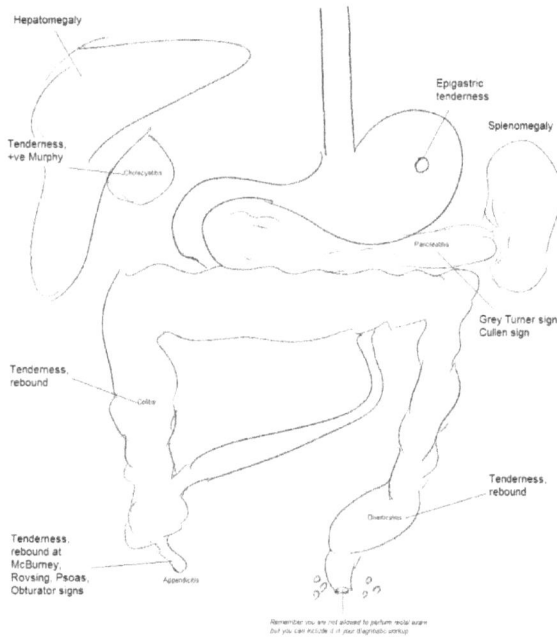

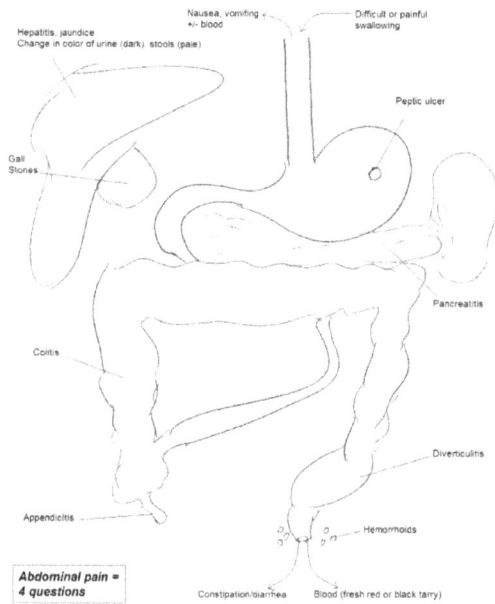

[11] Constipation = GI + general

DD	WU
1- Bowel obstruction	1- CBC with differential
2- Colon mass	2- serum electrolytes
3- Functional constipation	3- Rectal exam
4- Ileus	4- Colonoscopy
5- Ogilvie syndrome	5- CT scan of the abdomen & pelvis
6- Abdominal wall hernia	6- Serum CEA level
7- Hypothyroidism	7- Serum TSH, T3, T4

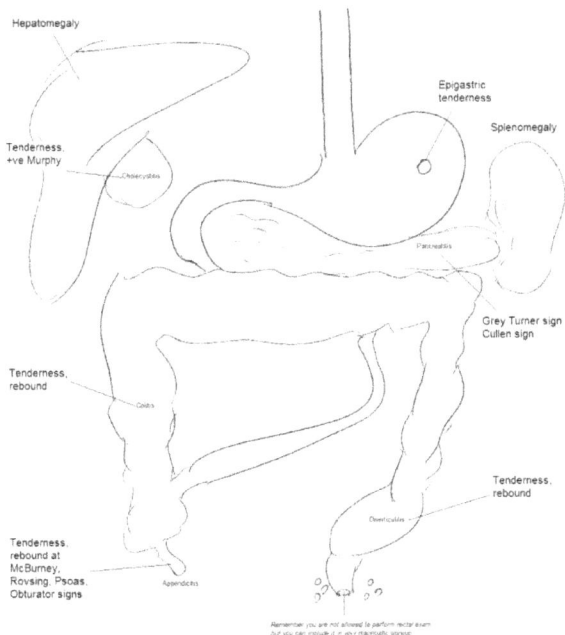

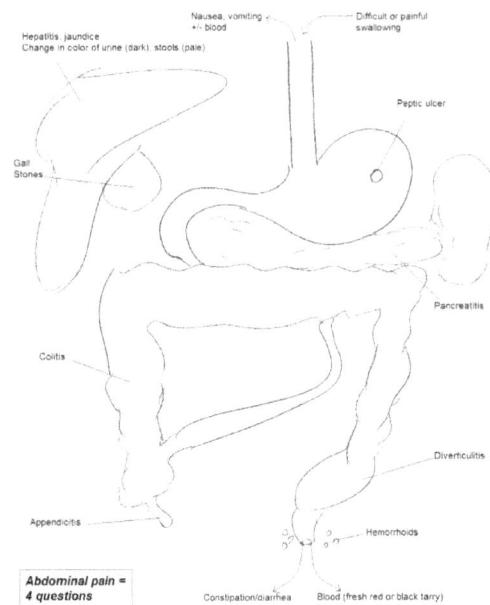

[12] Diarrhea = GI + general

DD	WU
1- Inflammatory bowel disease (Ulcerative colitis, Crohn's)	1- CBC with differential
2- Infectious colitis	2- serum electrolytes
3- Pseudomembranous colitis	3- Rectal exam
4- Celiac disease	4- Colonoscopy
5- Colorectal polyps/cancer	5- Stools for C-Diff toxin, ova, parasites, G stain, C&S
6- Cholelithiasis	6- Abdominal Ultrasound
7- VIPoma	7- CT scan of the abdomen & pelvis
	8- Serum CEA level

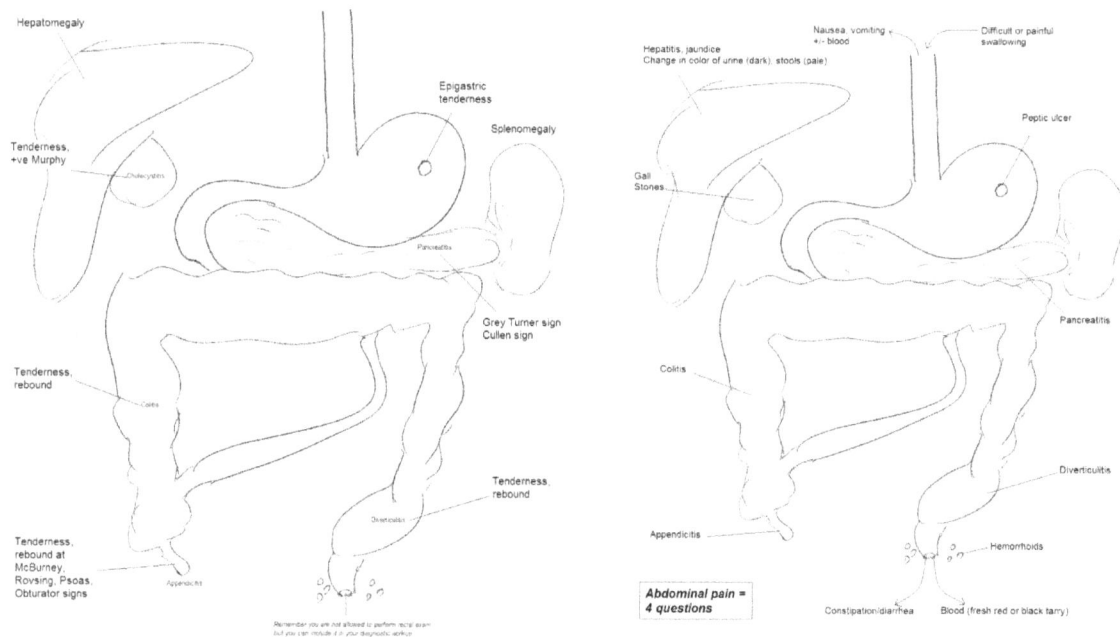

[13] Cough = Chest + general

DD	WU
1- Bronchitis	1- CBC with differential
2- Pneumonia	2- CXR (PA - Lat)
3- Tuberculosis	3- Sputum G. stain, culture & sensitivity
4- Asthma	4- Arterial blood gases (ABG)
5- CHF	5- O_2 saturation
6- Lung cancer	6- Chest CT scan
7- Bronchiectasis	7- Bronchoscopy
	8- PPD

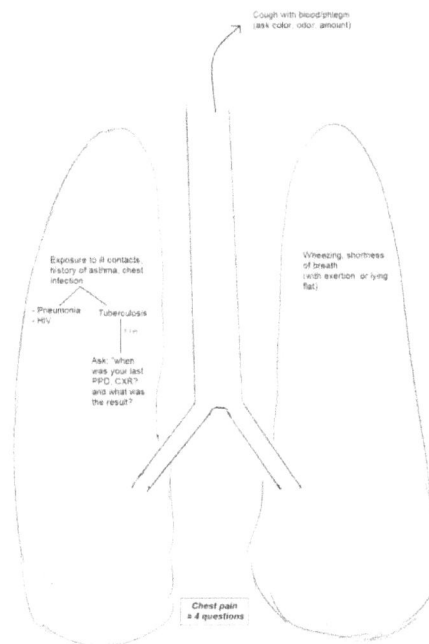

[14] Coughing up blood = Chest + general

DD	*WU*
1- Pulmonary embolism	1- CBC with differential
2- Pneumonia	2- Chest CT scan
3- Tuberculosis	3- PPD
4- Lung cancer	4- Bronchoscopy
5- Bronchiectasis	5- CXR (PA - Lat)
6- Bronchitis	6- Arterial blood gases (ABG)
7- CHF	7- O_2 saturation

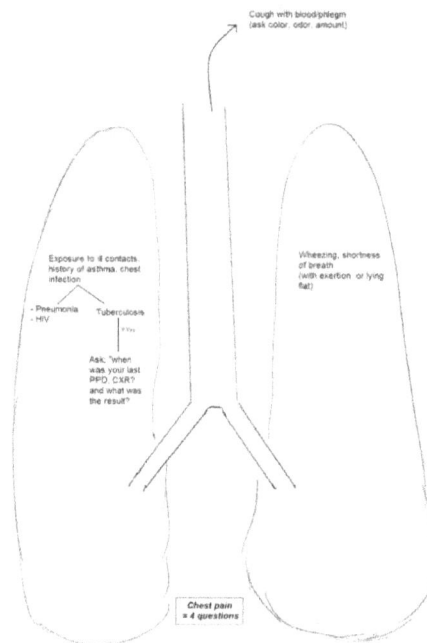

[15] **Wheezing in adults** = Chest + general

DD	WU
1- Asthma	1- CBC with differential
2- COPD	2- CXR (PA - Lat)
3- GERD	3- Chest CT scan
4- Bronchitis	4- Bronchoscopy
5- Bronchiectasis	5- O_2 saturation
6- Lung cancer	6- Arterial blood gases (ABG)
7- Pneumonia	7- Sputum G. stain, culture & sensitivity
	8- Upper GI endoscopy

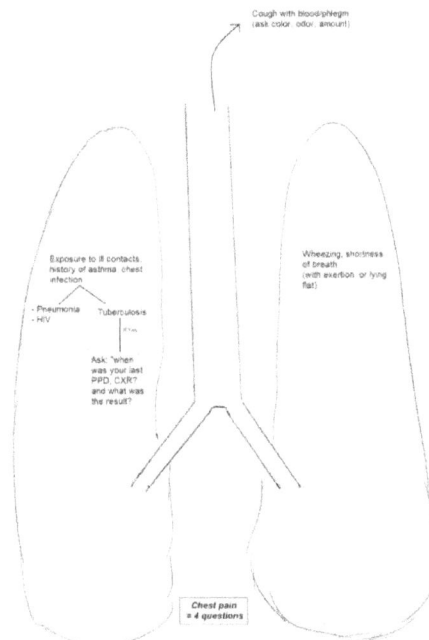

Cough with blood/phlegm
(ask color, odour, amount)

Exposure to ill contacts,
history of asthma, chest
infection

- Pneumonia   Tuberculosis
- HIV

Ask: "when
was your last
PPD, CXR?
and what was
the result?

Wheezing, shortness
of breath
(with exertion or lying
flat)

Chest pain
= 4 questions

[16] Wheezing in children = Chest + general

DD	WU
1- Enlarged tonsils or adenoids	1- CBC with differential
2- Asthma	2- CXR (PA - Lat)
3- Croup	3- Indirect laryngoscopy / Bronchoscopy
4- Tracheomalacia	4- Upper GI endoscopy
5- Pertussis	5- Chest CT scan
6- Bronchitis	6- Arterial blood gases (ABG)
7- Pneumonia	7- Sputum G. stain, culture & sensitivity
8- GERD	8- Upper GI endoscopy
	9- O_2 saturation

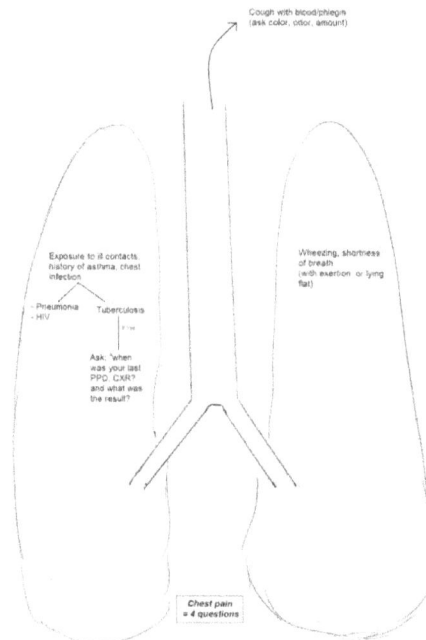

[17] Night sweats = Chest + Endocrine + general, don't forget to examine lymph nodes for lymphoma

DD	WU
1- Tuberculosis	1- PPD & CXR
2- Lymphoma	2- CT scan of chest & abdomen
3- Hyperthyroidism	3- Serum TSH, T3, T4
4- Pheochromocytoma	4- Serum catecholamines, urine metanephrines (VMA, HMA)
5- HIV	5- ELISA test for HIV
6- Menopause	6- CBC with differential
7- Idiopathic hyperhidrosis	7- ESR
8- Infection	8- Serum electrolytes
9- Hypoglysemia	9- Serum blood sugar

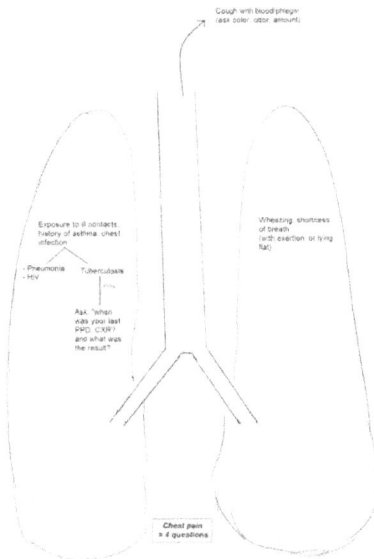

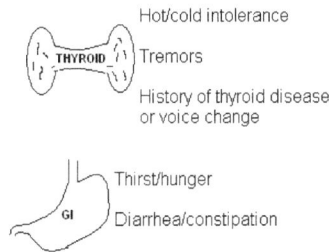

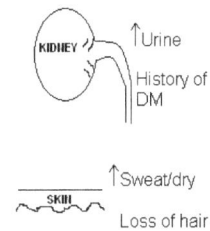

[18] Shortness of breath = Chest + cardiac + general

DD	WU
1- Asthma	1- CBC with differential
2- Pneumonia	2- CXR (PA - Lat)
3- cardiac (angina – infarct)	3- 12 lead EKG, cardiac enzymes
4- Pulmonary embolism	4- Chest CT scan
5- CHF	5- Arterial blood gases (ABG)
6- Bronchitis	6- O_2 saturation
7- Pneumothorax	7- Pulmonary function tests (PFT)

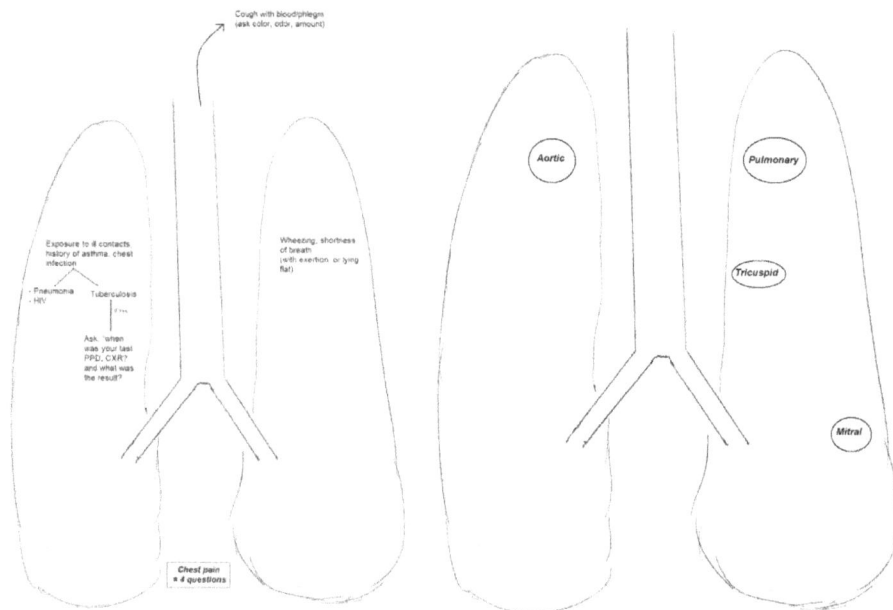

[19] Chest pain = Cardiac + chest + general

DD	WU
1- Myocardial infarction	1- 12 lead EKG, cardiac enzymes
2- Angina	2- Chest CT scan
3- Pulmonary embolism	3- CXR (PA - Lat)
4- Aortic dissection	4- Upper GI endoscopy
5- Pericarditis	5- Barium swallow
6- Costochondritis	6- Esophageal 24h PH monitoring
7- GERD	7- Esophageal manometry
8- Esophageal spasm	8- Upper GI endoscopy
9- Nutcracker esophagus	
10- Achalasia	

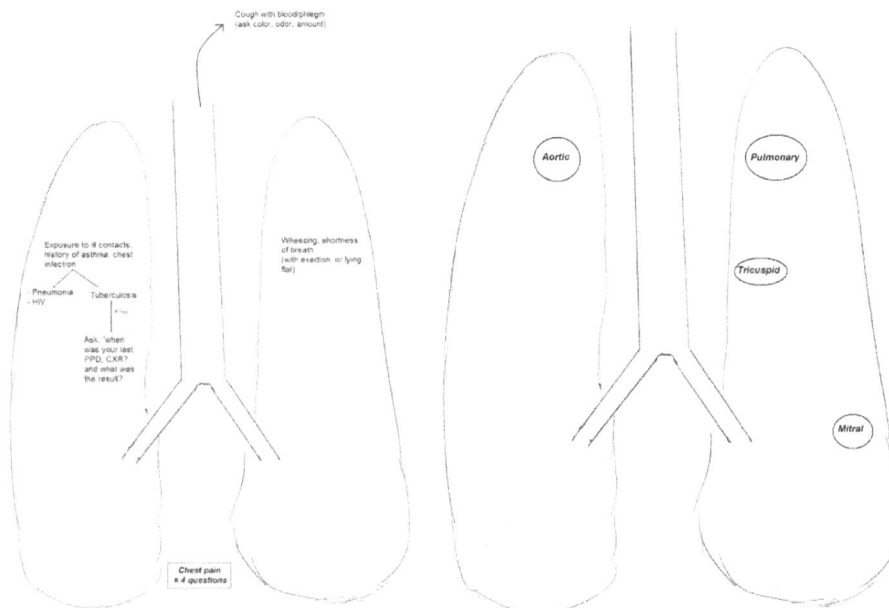

[20] **Palpitations** = Cardiac + general (and palpate the *thyroid*)

DD	WU
1- Cardiac dysrhythmia	1- 12 lead EKG, cardiac enzymes
2- Cardiac valvular lesion	2- Echocardiogram
3- Hyperthyroidism	3- Serum TSH, T3, T4
4- Hypoglycemia	4- Serum glucose level
5- Pheochromocytoma	5- Serum & Urine metanephrines
6- Electrolyte disturbance	6- Serum electrolytes

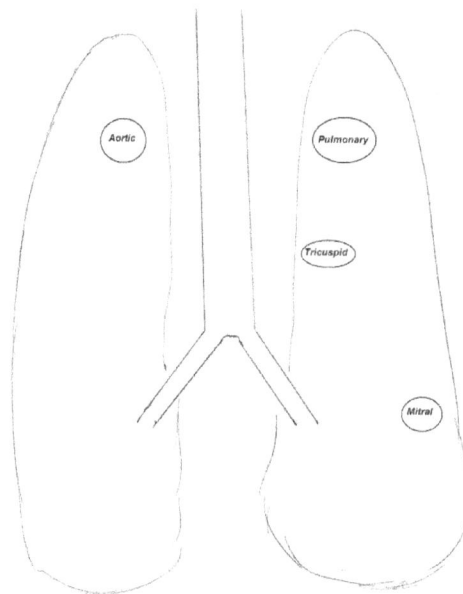

[21] Hypertension = Cardiac + general
- Measure the blood pressure in both arms, sitting and standing.
- Palpate pulses (carotid, radial, dorsalis pedis)

DD	WU
1- Essential hypertension	1- 12 lead EKG
2- Atherosclerosis	2- Serum lipid profile
3- Conn's syndrome	3- Serum renin and aldosterone level
4- Cushing's syndrome	4- Serum cortisol level
5- Pheochromocytoma	5- Serum and urine metanephrines
6- Renal artery stenosis	6- Ultrasound of the renal arteries
7- Obesity induced metabolic syndrome	7- Kidney functions
8- Coarctation of the aorta	8- Serum electrolytes

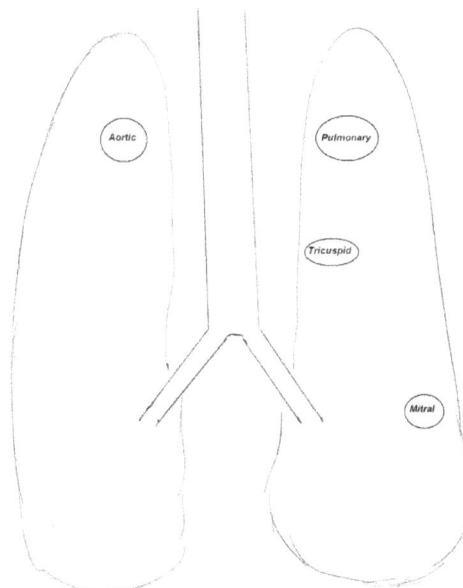

[22] Tremors = Neuro + general

DD	WU
1- Parkinson disease	1- Brain CT scan
2- Essential familial tremors	2- Brain MRI
3- Brain tumor	3- Serum TSH, T3, T4
4- Hyperthyroidism	4- Serum Ceruloplasmin
5- Wilson's disease	5- Serum copper
6- Peripheral neuropathy	6- Electromyogram
7- Carbon monoxide poisoning	7- serum carboxyhemoglobin
8- Huntington's disease	8- Genetic testing

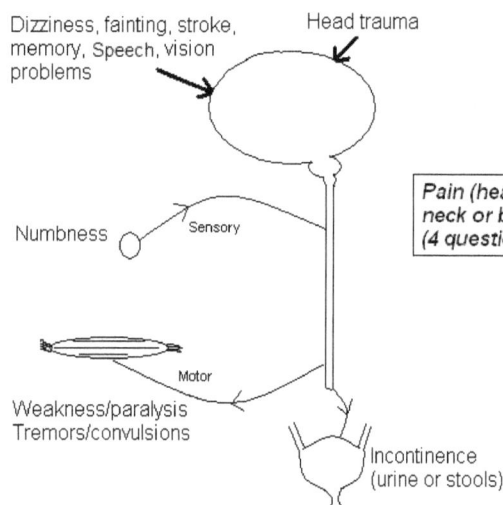

Dizziness, fainting, stroke, memory, Speech, vision problems

Head trauma

Numbness

Sensory

Pain (headache, neck or back) (4 questions)

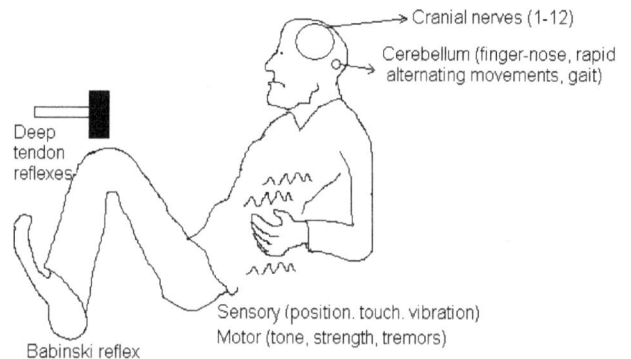

Motor

Weakness/paralysis
Tremors/convulsions

Incontinence (urine or stools)

Cranial nerves (1-12)

Cerebellum (finger-nose, rapid alternating movements, gait)

Deep tendon reflexes

Sensory (position. touch. vibration)
Motor (tone, strength, tremors)

Babinski reflex

[23] Limb weakness or paralysis = Neuro + general

DD	WU
1- Stroke	1- Head CT scan
2- Transient ischemic attack	2- Carotid duplex scan
3- Brain tumor	3- EKG
4- Trauma (head or spinal cord)	4- Echocardiogram
5- Carotid stenosis	5- MRI of the spine
6- Intracranial hemorrhage	6- MRI of the head
7- Meningitis	7- Lumbar puncture
8- Guillain-Barre syndrome	

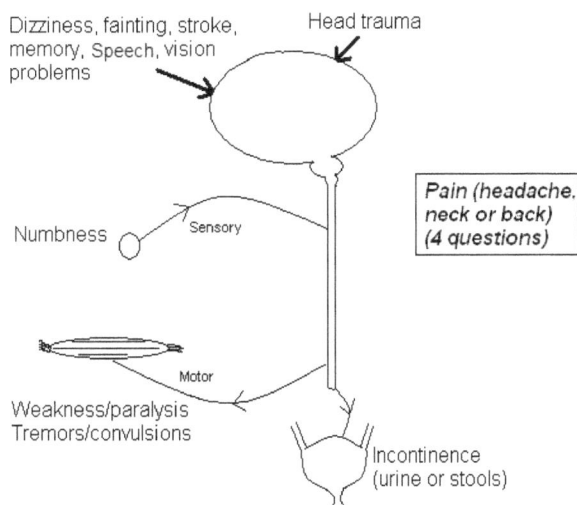

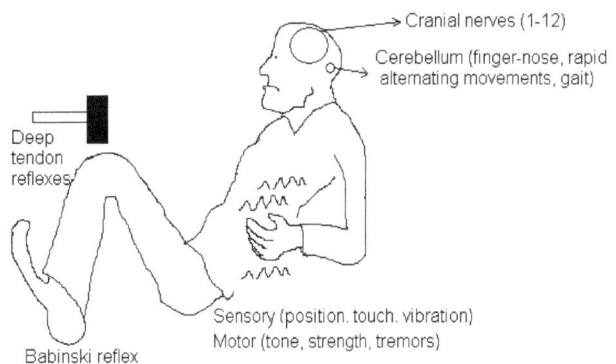

Dizziness, fainting, stroke, memory, Speech, vision problems

Head trauma

Numbness

Sensory

Pain (headache, neck or back) (4 questions)

Weakness/paralysis Tremors/convulsions

Motor

Incontinence (urine or stools)

Deep tendon reflexes

Cranial nerves (1-12)

Cerebellum (finger-nose, rapid alternating movements, gait)

Sensory (position. touch. vibration)
Motor (tone, strength, tremors)

Babinski reflex

[24] Tingling = Neuro + general
+ Add to exam: Tinel & Phalen tests

DD	WU
1- Carpal tunnel syndrome	1- Electromyography
2- Transient ischemic attack	2- Carotid duplex
3- Cervical radiculopathy	3- Cervical X-ray
4- Ulnar neuropathy	4- Serum blood sugar
5- Thoracic outlet syndrome	5- MRI of the cervical spine

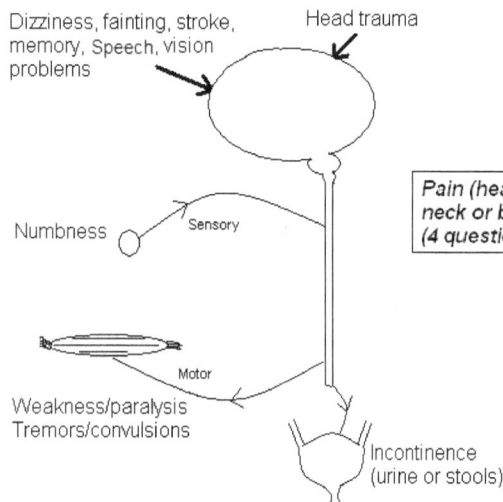

Dizziness, fainting, stroke, memory, Speech, vision problems

Head trauma

Numbness

Sensory

Pain (headache, neck or back) (4 questions)

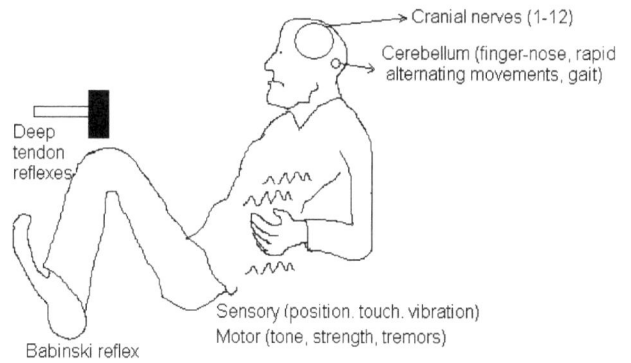

Weakness/paralysis
Tremors/convulsions

Motor

Incontinence (urine or stools)

Cranial nerves (1-12)

Cerebellum (finger-nose, rapid alternating movements, gait)

Deep tendon reflexes

Babinski reflex

Sensory (position. touch. vibration)
Motor (tone, strength, tremors)

[25] Dizziness or loss of consciousness = Neuro + general (& auscultate carotids for bruit)

DD	WU
1- TIA	1- Carotid duplex
2- CVA	2- CT scan of the head
3- Acoustic neuroma or brain tumor	3- MRI of the brain
4- Cardiac dysrhythmia	4- 12 lead EKG
5- Hyperthyroidism	5- Serum TSH, T3, T4
6- Hypoglycemia	6- Serum glucose level
7- Benign positional vertigo	7- Serum electrolytes
8- Anemia	8- CBC with differential
9- Orthostatic hypotension	

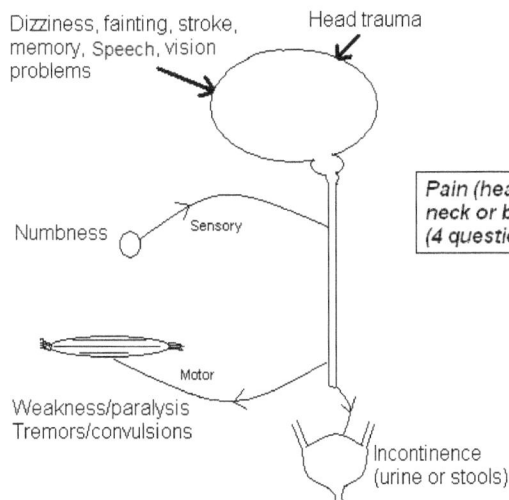

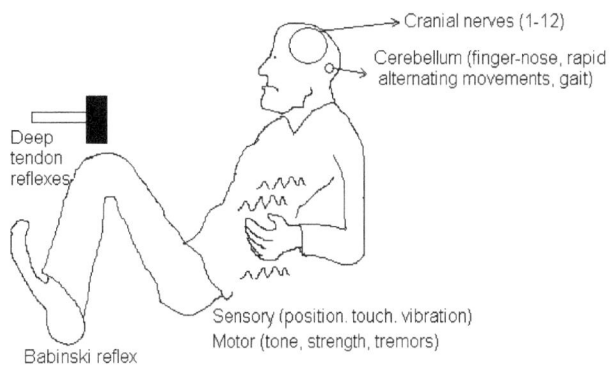

[26] Forgetfulness = Neuro + psychiatry + general

DD	WU
1- Alzheimer's disease	1- MRI brain
3- Depression	2- serum electrolytes
4- Subdural hematoma	5- CT brain
2- Hypothyroidism	4- Serum TSH, T3, T4
6- Vitamin B_{12} deficiency	6- Serum Vitamin B_{12} level
7- Folic acid deficiency	7- Serum folic acid level

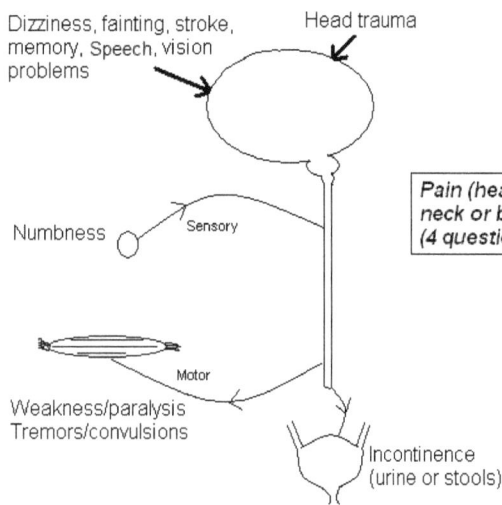

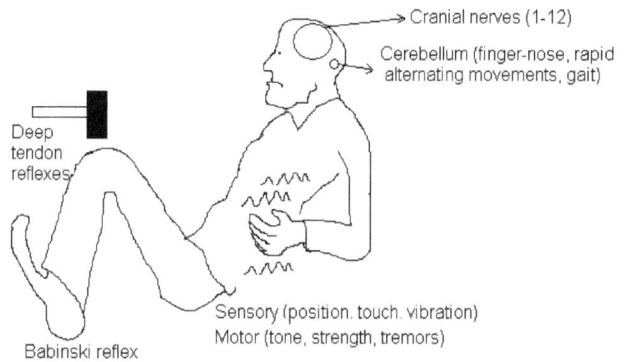

Dizziness, fainting, stroke, memory, Speech, vision problems

Head trauma

Numbness

Sensory

Pain (headache, neck or back) (4 questions)

Weakness/paralysis Tremors/convulsions

Motor

Incontinence (urine or stools)

Deep tendon reflexes

Cranial nerves (1-12)

Cerebellum (finger-nose, rapid alternating movements, gait)

Sensory (position. touch. vibration)
Motor (tone, strength, tremors)

Babinski reflex

[27] Headache = Neuro + general
+ Add to exam: neck looking for stiffness (meningitis)

DD	WU
1- Migraine	1- CT brain
2- Tension headache	2- CT of the facial sinuses
3- Cluster headache	3- CBC with differential
4- Hypertension	4- Spinal tap
5- Sinusitis	5- Temporal artery biopsy
6- Brain tumor	6- MRI brain
7- Meningitis	
8- Intracranial hemorrhage	

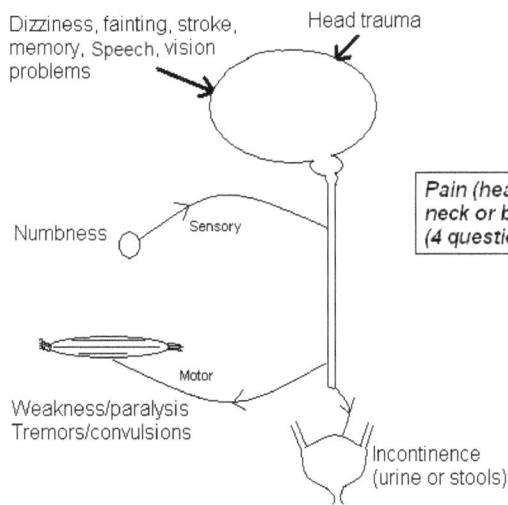

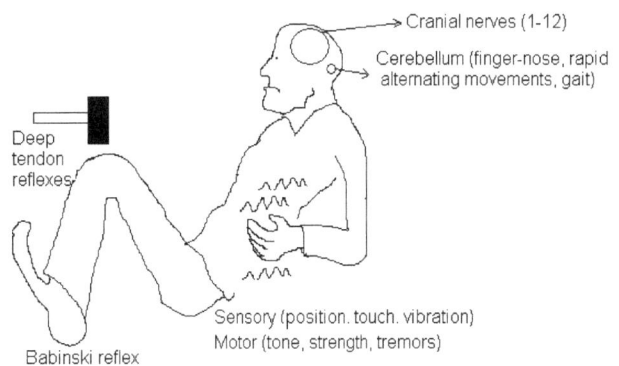

Dizziness, fainting, stroke, memory, Speech, vision problems

Head trauma

Numbness

Sensory

Pain (headache, neck or back) (4 questions)

Weakness/paralysis Tremors/convulsions

Motor

Incontinence (urine or stools)

Deep tendon reflexes

Cranial nerves (1-12)

Cerebellum (finger-nose, rapid alternating movements, gait)

Sensory (position. touch. vibration)
Motor (tone, strength, tremors)

Babinski reflex

[28] Breast Mass = Breast + general

DD	WU
1- Cyst	1- mammography
2- Abscess	2- Breast ultrasound
3- Fibroadenoma	3- Fine needle aspiration biopsy
4- Breast cancer	4- Breast biopsy
5- Galactocele	5- MRI
6- Fibrocystic disease	

Lumps & family history of breast cancer

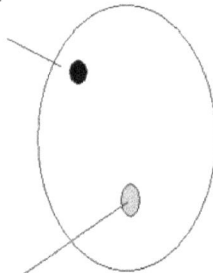

Nipple retraction, discharge, bleeding

Breast pain = 4 questions

+ MAM

1) mammogram
2) assymetry
3) monthly self exam

[29] Nipple Discharge = Breast + general

DD	WU
1- Galactorrhea	1- Mammography
2- Abscess	2- Breast ultrasound
3- Ductal ectasia	3- Serum & urine β-HCG level
4- Intraductal papilloma	4- Serum prolactin level
5- Intraductal carcinoma	5- Galactography
6- Fibrocystic disease	6- Ductoscopy

Lumps & family
history of breast
cancer

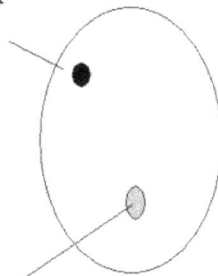

**Breast pain =
4 questions**

Nipple retraction,
discharge, bleeding

+ MAM
1) mammogram
2) assymetry
3) monthly self exam

[30] Blurred vision = Eyes + general

DD	WU
1- Error of refraction	1- Retinoscopy
2- Cataract	2- Slit lamp examination
3- Glaucoma	3- Tonometry
4- Diabetic retinopathy	4- Serum blood sugar
5- Macular degeneration	5- Indirect fundus examination
6- Optic neuritis	6- Ultrasound
7- Retinal detachment	
8- Uveitis	
9- Hypertension	

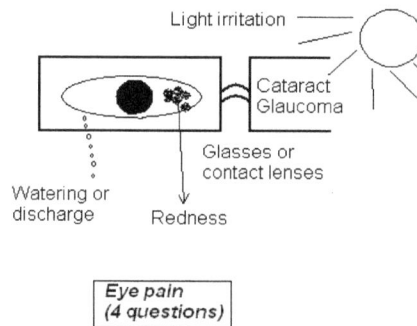

Double vision

Blurred vision

Light irritation

Cataract
Glaucoma

Glasses or contact lenses

Watering or discharge

Redness

Eye pain
(4 questions)

[31] Eye discharge = Eyes + general

DD	WU
1- Acute conjuctivitis	1- G stain, culture & sensitivity
2- Stye	2- Slit lamp examination
3- Chalazion	3- Ultrasound
4- Corneal abrasion	4- Fluorescein test
5- Foreign body	5- CBC with differential

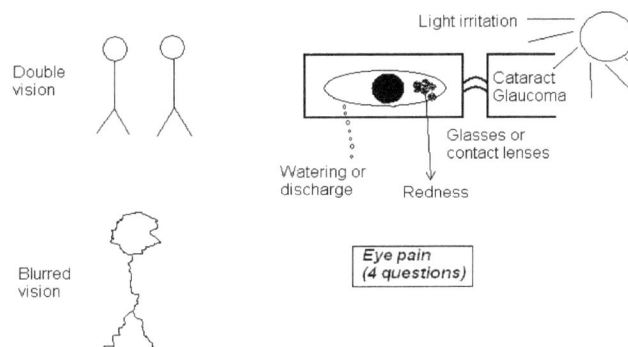

Double vision

Blurred vision

Light irritation

Cataract
Glaucoma

Glasses or
contact lenses

Watering or
discharge

Redness

Eye pain
(4 questions)

[32] Decreased hearing = Ears + general

DD	WU
1- Cerumen impaction	1- CBC
2- Labyrinthitis	2- Audiogram
3- Perforation of tympanic membrane	3- CT brain
4- Presbycusis	4- Serum electrolytes
5- Otitis media	5- Liver functions
6- Acoustic neuroma	

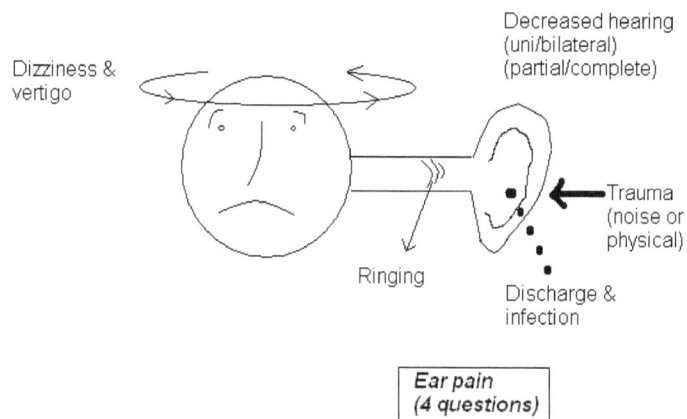

Dizziness & vertigo

Decreased hearing (uni/bilateral) (partial/complete)

Trauma (noise or physical)

Ringing

Discharge & infection

Ear pain (4 questions)

[33] Difficulty swallowing = Throat + general

DD	WU
1- Achalasia	1- Upper GI endoscopy
2- Esophageal stricture	2- Barium swallow
3- Esophageal cancer	3- Esophageal 24h PH monitoring
4- Esophageal spasm	4- Esophageal manometry
5- Schatzki ring	5- Biopsy of esophageal mass
6- Nutcracker esophagus	
7- Scleroderma	

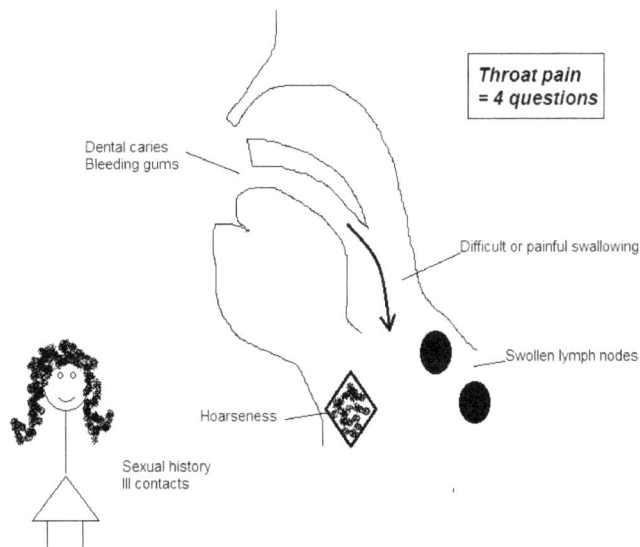

Throat pain = 4 questions

Dental caries
Bleeding gums

Difficult or painful swallowing

Swollen lymph nodes

Hoarseness

Sexual history
Ill contacts

[34] Sore throat & jaundice = Throat + GI + general

DD	WU
1- Tonsillitis	1- CBC with differential
2- Pharyngitis	2- Monospot test
3- Infectious mononucleosis	3- Blood smear
4- Hepatitis	4- Liver functions
5- Rotor syndrome	5- Total, direct, indirect bilirubin
6- Dubin-Johnson syndrome	6- Abdominal ultrasound
7- Gilbert syndrome	

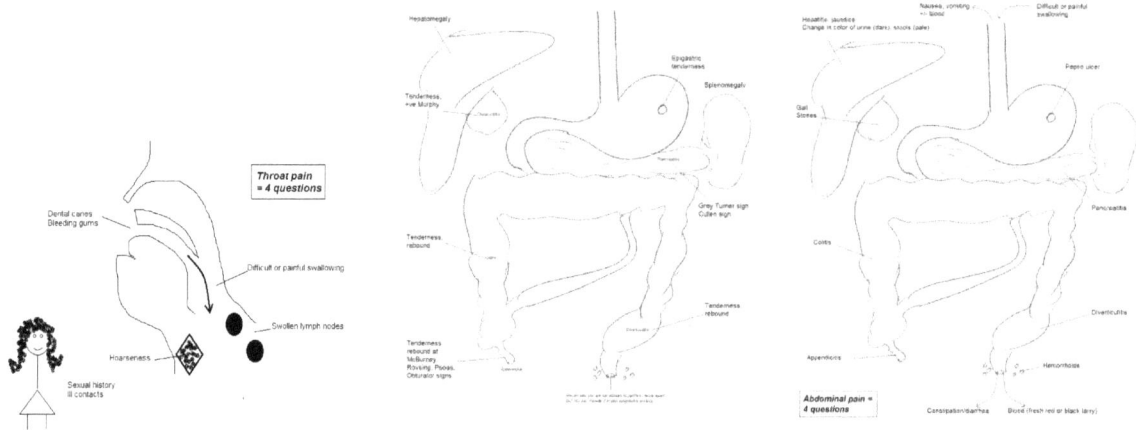

[35] Fatigue, weight change = Endocrine + psychiatry + general

DD	WU
1- Hypothyroidism	1- CBC with differential
2- Depression	2- Serum electrolytes
3- Anemia	3- Liver functions eg Albumin, ALT, AST
4- Infectious mononucleosis	4- Serum TSH, T3, T4
5- Congestive heart failure	5- Cholesterol level
6- HIV	6- CXR
7- Cancer	7- ELISA test for HIV
8- Tuberculosis	8- PPD skin test
9- Chronic fatigue syndrome	9- Echocardiogram
10- Diabetes	10- Serum blood sugar
11- Sleep apnea	11- Sleep study

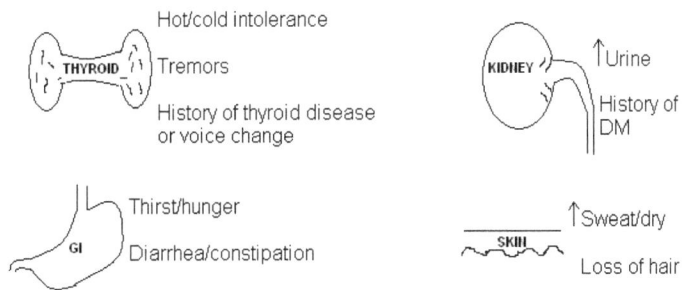

THYROID — Hot/cold intolerance
Tremors
History of thyroid disease or voice change

KIDNEY — ↑Urine
History of DM

GI — Thirst/hunger
Diarrhea/constipation

SKIN — ↑Sweat/dry
Loss of hair

[36] Introvert Patient = Psychiatry + general

DD	WU
1- Depression	1- CBC with differential
2- Bipolar disorder	2- Serum electrolytes
3- Hypothyroidism	3- TSH, T3, T4
4- Brain tumor	4- CT brain
5- Chronic fatigue syndrome	5- MRI of the brain

[37] Alcoholism = Psychiatry + general + palpate liver for hepatomegaly
Add the CAGE questionnaire to HPI:

C → Did you try to **C**ut down on drinking?

A → Are people **A**nnoying you because of drinking?

G → Do you feel **G**uilty about drinking?

E → Do you need an **E**ye opener (drink early in the day)?

DD	WU
1- Depression	1- CBC with differential
2- Anxiety	2- Serum electrolytes
3- Attention deficit disorder	3- Liver function tests (ALT, AST, GGT)
4- Schizophrenia	4- Liver ultrasound
5- Antisocial disorder	5- Upper GI endoscopy for varices, gastritis, ulcers

[38] Obesity = Psychiatric + endocrine + general
 Also ask about: emotional & binge eating.
Add to examination: Body Mass Index (BMI)
[BMI = weight (lbs) x 703 / height (inches) ^{2}]

DD	WU
1- Hypothyroidism	1- TSH, T3, T4
2- Cushing syndrome	2- Serum ACTH
3- Corticostroid intake	3- Serum cortisol level
4- Polycystic ovary	4- Pelvic ultrasound
5- Insulinoma	5- Serum insulin level
6- Oral contraceptives	6- Psychology evaluaton
7- Eating disorder (e.g. bulimia, binge eating)	7- Genetic testing
8- Down syndrome	
9- Familial obesity	

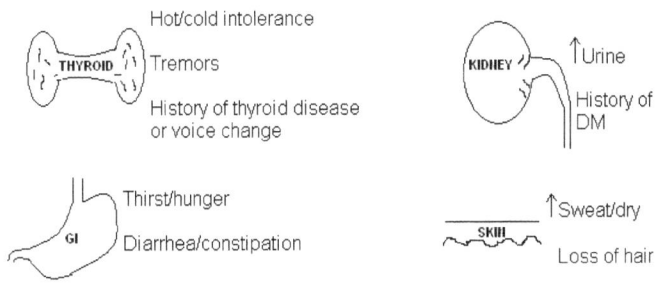

THYROID — Hot/cold intolerance
Tremors
History of thyroid disease or voice change

KIDNEY — ↑Urine
History of DM

GI — Thirst/hunger
Diarrhea/constipation

SKIN — ↑Sweat/dry
Loss of hair

[39] Joint pain e.g. knee = Musculoskeletal + general

DD	WU
1- Osteoarthritis	1- CBC with differential
2- Rheumatoid arthritis	2- Joint X-ray
3- Septic arthritis	3- Joint aspiration, G. stain, culture & sensitivity
4- Trauma	4- MRI of the joint
5- Gout	5- Serum uric acid
6- Collagen vascular disease	6- Urine analysis for crystals
7- Bone cancer	7- Erythrocyte sedimentation rate (ESR)
8- Lyme disease	8- ELISA & Western Blot
9- Osteoporosis	9- DEXA scan
10- Joint dislocation	
11- Tendonitis	

[40] Back pain = Musculoskeletal + general

DD	WU
1- Back strain	1- Lumbar X-ray
2- Sciatica	2- Lumbar MRI
3- Osteorthritis	3- Serum estrogen level
4- Prolapsed disc	4- Serum calcium, alkaline phosphatase
5- Ankylosing spondylitis	5- human leukocyte antigen-B27 assay
6- Osteoporosis	6- DEXA scan
7- Metastases (e.g. prostate cancer)	7- Rectal exam for prostate cancer, Serum PSA

[41] Upper extremity pain = Musculoskeletal + general

DD	WU
1- Thoracic outlet syndrome	1- Nerve conduction test
2- Carpal tunnel syndrome	2- CBC with differential
3- Cervical disc disease	3- MRI of cervical spine
4- Referred pain (angina, infarct)	4- 12 lead EKG
5- Trauma	5- Upper extremity X-ray
6- Tenosynovitis	6- ESR

[42] Shoulder pain = Musculoskeletal + general

DD	WU
1- Shoulder dislocation	1- Shoulder X-ray (2 views)
2- Bursitis or tendonitis	2- Shoulder MRI
3- Trauma	3- Serum rheumatoid factor, ANA
4- Ligament sprain	4- CBC with differential
5- Osteoarthritis	5- ESR

[43] Leg pain = Vascular + general

DD	WU
1- Acute arterial ischemia	1- Ultrasound of the lower extremities
2- Chronic claudication	2- Ankle brachial index (ABI)
3- DVT	3- D-dimer
4- Muscle hematoma	4- CBC with differential
5- Neuropathy (e.g. diabetic)	5- Serum glucose, Hgb A1C
6- Trauma	6- X-ray
7- Arthropathy	7- MRI
8- Fibromyalgia	8- Serum PT, PTT, INR
9- Cellulitis	

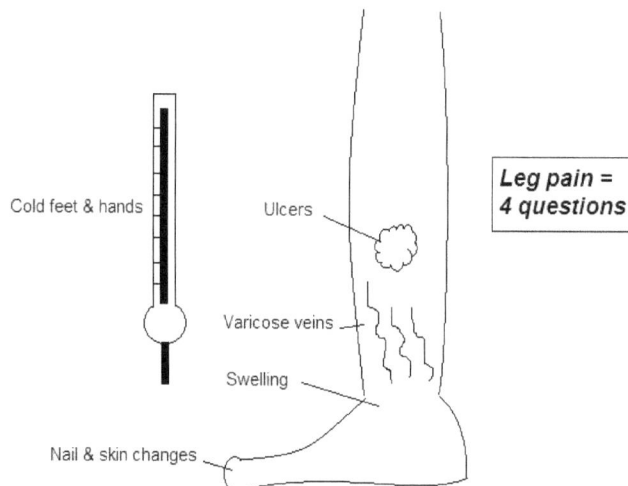

Cold feet & hands

Ulcers

Varicose veins

Swelling

Nail & skin changes

Leg pain =
4 questions

[44] Burning with urination = Urology + general

DD	WU
1- Urinary tract infection	1- CBC with differential
2- Pyelonephritis	2- Pelvic exam
3- Kidney stones	3- CT abdomen/pelvis without contrast
4- STD	4- Urine analysis, G stain, culture & sensitivity
5- Prostatitis	5- Rectal exam for prostatitis

[45] Difficulty urinating = Urology + general

DD	WU
1- Benign prostatic hypertrophy	1- CBC with differential
2- Cystitis	2- Urine analysis, G stain, culture & sensitivity
3- Kidney stones	3- Kidney function tests
4- Neurogenic bladder	4- PSA level
5- Malignancy	5- Rectal exam
6- Stricture	6- Intravenous pyelography
7- Prostatitis	7- Cystoscopy

[46] Hot flashes = Female genital system + general

DD	WU
1- Menopause	1- Serum estrogen, progesterone, FSH, LH
2- Hyperthyroidism	2- Serum TSH, T3, T4
3- Chronic fatigue syndrome	3- CBC with differential
4- Uterine or ovarian cancer	4- Pap smear
5- Factitious disorder	5- Serum electrolytes

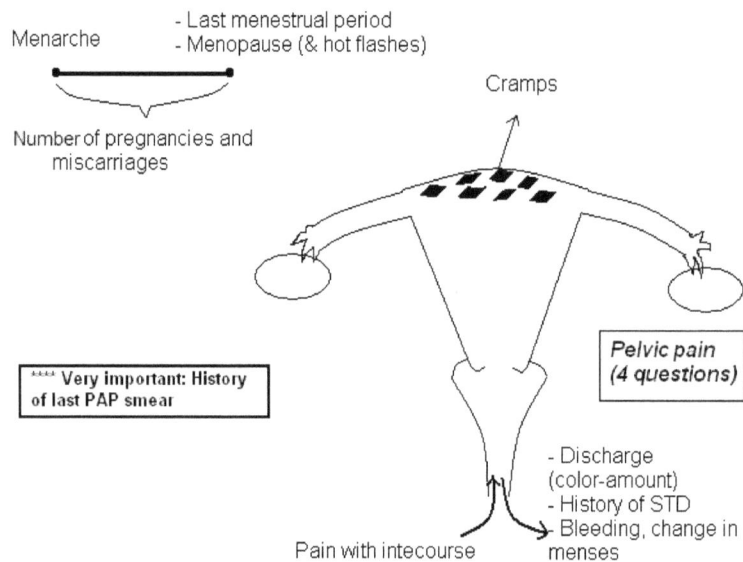

Menarche

- Last menestrual period
- Menopause (& hot flashes)

Number of pregnancies and miscarriages

Cramps

Pelvic pain (4 questions)

**** Very important: History of last PAP smear

- Discharge (color-amount)
- History of STD
- Bleeding, change in menses

Pain with intecourse

[47] Vaginal bleeding = Female genital system + general ......Remember that you're not allowed to do genital exam, but you need to examine the abdomen.

DD	*WU*
1- Pregnancy (intrauterine or ectopic)	1- Pelvic examination
2- Abortion	2- Serum & urine β-HCG level
3- Polycystic ovary	3- Abdominal & pelvic ultrasound
4- Uterine fibroids	4- CBC with differential
5- Hydatiform mole	5- TSH, T3, T4
6- Placental abruption	
7- Hypo or hyperthyroidism	

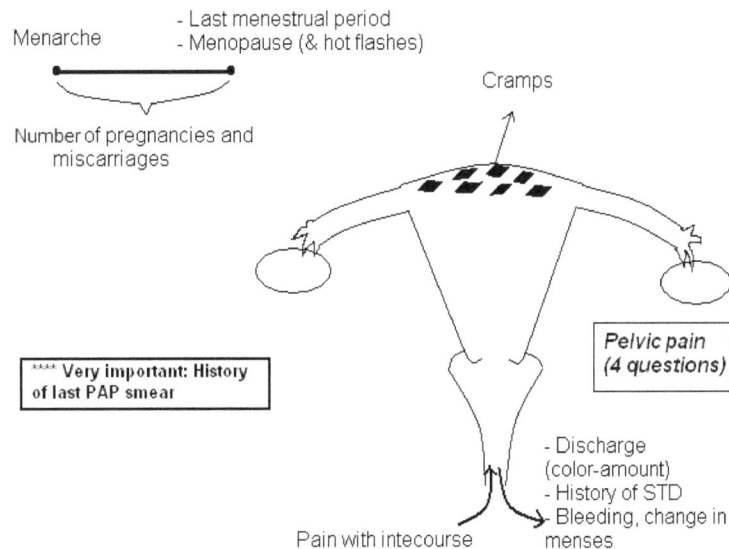

Menarche

- Last menestrual period
- Menopause (& hot flashes)

Number of pregnancies and miscarriages

Cramps

**** Very important: History of last PAP smear

Pelvic pain (4 questions)

Pain with intecourse

- Discharge (color-amount)
- History of STD
- Bleeding, change in menses

[48] Child with fever

In HPI think GI, chest, UTI, immunization

Examine: Throat, ear, neck for lymphadenopathy, lungs, abdomen

DD	WU
1- Otitis media	1- CBC with differential
2- Sinusitis	2- Facial sinus X-ray
3- Chest infection	3- CXR
4- Gastroenteritis	4- Abdominal X-ray
5- Tonsillitis	5- Stools for G stain, culture & sensitivity, ova & parasites
6- Lymphadenitis	

Conclusion

- Please remember that you need to focus on establishing a methodical way of thinking rather than memorizing each individual case.

- This book is designed for a wrap up review before the exam; it's short, concise and straight to the point.

- Try to forget that the "patients" in the exam room are your "examiners". Assume the role of the physician evaluating real patients.

- At the end, I would like to sincerely wish you the best of luck in the exam and in all your future endeavors.

- Please feel free to contact me if you have any questions or to provide feedback about this book. My email is: ehabakkary@yahoo.com

Ehab Akkary MD FASMBS
Director of Bariatric & Advanced Laparoscopic Surgery
Preston Memorial Hospital, Kingwood, WV, USA

References:

1- www.usmle.org

2- Tazkarji M. Abdominal Pain Among Older Adults. Geriatrics and Aging. 2008;11(7):410-415.

3- Jones R, Claridge J. Acute abdomen. Sabiston Textbook of Surgery: The Biologic Basis of Modern Surgical Practice, 17th edition. Philadelphia: Elsevier; 2004:1219-38.

4- Adam A, Benjamin I. Assessment of diagnostic techniques for biliary obstruction and liver masses. Surgery of the Liver and Biliary Tract. New York: Churchill Livingstone, 1994; 401-13.

5- Chang A, Asher M. A review of cough in children. J Asthma 2001; 38:299-309.

6- Braman S, Corrao W. Chronic cough. Prim Care 1985;12:217-25.

7- Poe R, Israel R, Utell M, Hall W. Chronic cough: bronchoscopy or pulmonary function testing? Am Rev Respir Dis 1982;126:160-162.

8- Bidwell J, Pachner R. Hemoptysis: Diagnosis and Management. Am Fam Physician. 2005; 72(7):1253-1260.

9- Howard W. Differential Diagnosis of Wheezing in Children. Pediatrics in Review. 1980;1:239-244.

10- Woollard M, Greaves I. Shortness of Breath M. Emerg Med J. 2004;21:341–350.

11- Klinkman M, Stevens D, Gorenflo D. Episodes of care for chest pain: a preliminary report from MIRNET. Michigan Research Network. J Fam Pract 1994;38:345.

12- Phan A, Shufelt C, Merz C. Persistent chest pain and no obstructive coronary artery disease. JAMA 2009; 301:1468.

13- Akkary E, Caranasos T, Martin L. Current Surgical Treatment of Gastro-oesophageal Reflux Disease. European Gastroenterology & Hepatology Review, 2009;5:7-10

14- Akkary E, Azam J, Caranasos T, Kupec J, Kusti M. Achalasia. US Gastroenterology & Hepatology Review. 2010;6:11–6

15- Abbott A. Diagnostic Approach to Palpitations. Am Fam Physician. 2005;71(4):743-750.

16- Crawford P, Zimmerman E. Differentiation and diagnosis of tremor. Am Fam Physician. 201;83(6):697-702.

17- LoVecchio F, Jacobson S. Appraoch to generalized weakness and peripheral neuromuscular disease. Emerg Med Clin N Am 1997;15(3): 605-623.

18- Brinton L, DeVesa S. Etiology and pathogenesis of breast cancer, in Harris JR, Lippman ME, Morrow M, et al (eds): Diseases of the Breast. Philadelphia, Lippincott-Raven, 1996: 159-168.

19- Garber J, Smith B. Management of the high-risk and the concerned patient, in Harris JR, Lippman ME, Morrow M, et al (eds): Diseases of the Breast. Philadelphia, Lippincott-Raven, 1996: 323-334.

20- Falkenberry S. Nipple Discharge. Obstet Gynecol Clin North Am. 2002;29(1):21-29.

21- Hellmann D. Arthritis & Musculoskeletal Disorders.Current Medical Diagnosis and Treatment 2004. 43rd ed. New York: McGraw-Hill, 2003;778-832.

22- Beers M. Osteoarthritis. The Merck Manual of Medical Information. 2nd Home ed. New York: Simon and Schuster, 2003. 367-370.

23- Jacobs-Kosmin D et al. Osteoporosis. eMedicine. Medscape. 23 Dec. 2008.

24- Carretero O, Oparil S. Essential hypertension. Part I: definition and etiology. Circulation 2000:101(3): 329–335.

25- www.bettermedicine.com/article/leg-pain.

26- Isaacson J, Vora N, Milton S. Differential Diagnosis and Treatment of Hearing Loss. Am Fam Physician. 2003;68(6):1125-1132.

27- http://www.entnet.org/HealthInformation/soreThroats.cfm.

28- Knopman D, DeKosky S, Cummings J et al. Practice parameter: diagnosis of dementia (an evidence-based review). Report of the Quality Standards Subcommittee of the American Academy of Neurology. Neurology 2001;56:1143– 1153.

29- Talwalker S. The cardinal features of cognitive and noncognitive dysfunction and the differential efficacy of tacrine in Alzheimer disease patient. J Biopharm Stat 1996;6:443– 456.

30- Detsky M, McDonald D, Baerlocher M, Tomlinson G, McCrory D, Booth C. Does this patient with headache have a migraine or need neuroimaging? JAMA 2006;296:1274–1283.

31- Collins R. Blurred vision, blindness and stocomata. Chapter in "Differential Diagnosis in Primary Care" 2007;67 Lippincott Williams & Wilkins.

32- http://www.wrongdiagnosis.com/sym/burning_during_urination.htm.

33- Sharp L, Lipsky M. Screening for depression across the lifespan: a review of measures for use in primary care settings. Am Fam Physician. 2002;66 (6): 1001–1008.

34- González H, Vega W, Williams D, Tarraf W, West B, Neighbors H. Depression Care in the United States: Too Little for Too Few. Archives of General Psychiatry. 2010;67(1): 37–46.

35- Patel A, Ogle A. Diagnosis and Management of Acute Low Back Pain. Am Fam Physician. 2000;61:1779-1786,1789-1790.

36- Futterweit W, Ginzburg S, Goodman N, Kleerekoper M et al. Endocrinologists' Guidelines for the Diagnosis and Treatment of Menopause: Nonhormonal Therapy for Menopause. Endocrine Practice. 2006;12(3):315-337.

37- Neal S, Fields K. Peripheral nerve entrapment and injury in the upper extremity. Am Fam Physician. 2010;81(2):147-155.

38- http://www.emedicinehealth.com/night_sweats/article_em.htm

39- Ewing J. Detecting Alcoholism: The CAGE Questionnaire. JAMA. 1984;252: 1905-1907.

40- Claessens E, Cowell C. Dysfunctional uterine bleeding in the adolescent. Pediatric Clinics of North America. 1981;28: 369-378.

41- Mitan L, Slap G. Dysfunctional uterine bleeding. In Adolescent health care: A practical guide (5th ed.) Philadelphia: Lippincott Williams & Wilkins. 2008: 687-690.

42- White R. Shoulder pain. West J Med. 1982;137(4): 340–345

43- Overweight and Obesity. National Center for Chronic Disease Prevention and Health Promotion. Centers for Disease Control and Prevention. 27 Aug. 2009 http://www.cdc.gov/obesity/index.html.

44- http://www.emedicinehealth.com/fatigue/page2_em.htm.